Lecture Notes in Applied and Computational Mechanics

Volume 17

Springer

Berlin
Heidelberg
New York
Hong Kong
London
Milan
Paris
Tokyo

Structural Reliability

Statistical Learning Perspectives

Jorge E. Hurtado

Springer

Dr. Jorge E. Hurtado
Universidad Nacional de Colombia
Civil Engineering
Apartado 127
Manizales
Colombia

With 130 Figures

Library of Congress Control Number: 2004105238

ISBN 3-540-21963-3 Springer Verlag Berlin Heidelberg New York

Springer-Verlag is a part of Springer Science+Business Media

springeronline.com

Printed in Germany

Cover design: design & production GmbH, Heidelberg
Typesetting: Digital data supplied by authors

Printed on acid-free paper 62/3020Rw -5 4 3 2 1 0

To Eugenia

Preface

The last decades have witnessed the development of methods for solving structural reliability problems, which emerged from the efforts of numerous researchers all over the world. For the specific and most common problem of determining the probability of failure of a structural system in which the limit state function $g(\boldsymbol{x}) = 0$ is only implicitly known, the proposed methods can be grouped into two main categories:

- Methods based on the Taylor expansion of the performance function $g(\boldsymbol{x})$ about the most likely failure point (the design point), which is determined in the solution process. These methods are known as FORM and SORM (First- and Second Order Reliability Methods, respectively).
- Monte Carlo methods, which require repeated calls of the numerical (normally finite element) solver of the structural model using a random realization of the basic variable set $\boldsymbol{x}$ each time.

In the first category of methods only SORM can be considered of a wide applicability. However, it requires the knowledge of the first and second derivatives of the performance function, whose calculation in several dimensions either implies a high computational effort when faced with finite difference techniques or special programs when using perturbation techniques, which nevertheless require the use of large matrices in their computations. In order to simplify this task, use has been proposed of techniques that can be regarded as variants of the Response Surface Method. However, according to some recent research and the theoretical discussion done in this book, the results offered by the method are characterized by a high dependence on the sampling plan.

On the other hand the Monte Carlo method is characterized by its entire generality. Nonetheless, it is a well known fact that it requires a large computational labor. With respect to FORM and SORM it requires random samples instead of deterministic ones and its scaling with dimensionality is rather weak. Indeed, together with its generality, this later characteristic is a positive advantage. However, it remains the problem of its computational cost. This is not only due to the fact that it requires the many calls of the numerical solver for the assessment of any statistical quantity but also that in using it this quantity becomes a random variable.

In order to diminish the computing labor of Monte Carlo methods, research efforts oriented to a reduction of the number of samples and the spread of the estimates obtained at each simulation epoch have been exerted in the last years. Sometimes, however, these developments are accompanied by a sacrifice of one or more of the attractive properties of the basic Monte Carlo method. For instance, some of these variance-reduction methods require an exploration for locating the design point or the estimation of an optimal density function, which is a task seriously affected by the curse of dimensionality. Anyhow, these research efforts evidence the need of reducing the very high cost of Monte Carlo simulation especially for the case of large structures. An important step in this direction was made in the eighties with the application of the Response Surface Method developed in the field of Design of Experiments. Its basic idea is to calculate a simple function of the basic variables $\hat{g}(\boldsymbol{x})$ that could act as a surrogate of the finite element solver for estimating the performance function. The surrogate is then used for most of the samples needed in the entire Monte Carlo simulation programme. To such purpose, a experimental plan of samples is carried out for calculating the coefficients of the surface using standard algebraic or least-square techniques. The surrogate so obtained can also be used in connection to FORM and SORM methods.

These developments opened a new area of structural reliability research that can be labeled as the investigation on solver surrogate methods. Note that taking into account the refinement and sophistication of finite element models, the practical application of Monte Carlo simulation could be greatly favored by the availability of such a substitute to the actual deterministic solution, since only some finite element solver calls would be needed for building the surrogate. In the last years some important developments in this direction have occurred. First, some serious deficiencies of the Response Surface Method that make it unreliable for practical use have been highlighted. Second, the development of Statistical Learning Theory and algorithms, such as Neural Networks and Support Vector Machines, afforded new ways of obtaining the surrogate of the performance function. Third, attention has been called by the author to the fact that the structural reliability problem can be treated as a statistical classification task determined by the limit state function, and to this purpose some modern pattern recognition algorithms are especially useful.

In this context, the present monograph deals mainly with the problem of building such surrogates with the instruments provided by modern Statistical Learning Theory. Besides, some traditional reliability methods are examined under the light of such theory in order to shed light on some difficulties that are found in their application. The monograph, however, is not only oriented to propose new methods for solving the reliability problem but also to examine some important issues on the management of samples in Monte Carlo simulation for structural reliability analysis purposes. Generally speaking, the monograph has been inspired by two basic ideas: First, that the application of *any* structural reliability method would be benefited by the availability of a solver surrogate, under the condition that it grants a good performance when used with samples not employed in its construction. Second, that with the refinement of finite element meshes and constitutive models, engineers cannot indulge the luxury of generating samples without learning something from them. The development of several adaptive Monte Carlo techniques in structural reliability evidences this need. To these purposes, Statistical Learning Theory provides powerful algorithms and elegant answers that are worth examination from the point of view of structural reliability and adapted to its specific needs.

The book is organized as follows. Chapter 1 is devoted to a discussion on structural reliability methods, intended to highlight the relevance of building flexible and adaptive solver surrogates of finite element solvers that overcome the limitations of the Response Surface Method. Chapter 2 presents the fundamental concepts of Statistical Learning allowing an examination of learning techniques. The issues examined are the risk minimization, the inductive principles, the Vapnik-Chervonenkis dimension, the structure of learning models, the sample complexity, the curse of dimensionality and the problems of probability density estimation. The chapter is closed with a discussion on the criteria for selecting a learning method to solve structural reliability problems. In Chapter 3 the important themes of dimensionality reduction and data compression, which are important for processing the samples arising from a Monte Carlo simulation, are treated. The tools examined are the classical Principal Component Analysis, the nonlinear Kernel PCA, the Karhunen-Loève expansion, the wavelet transform, Vector Quantization and the Expectation-Maximization algorithm.

One of the main purposes of the monograph is to examine the treatment of the structural reliability problem as a pattern recognition (or classification) task. For this reason this subject is studied in three chapters. Chapter 4 starts with a discussion on the applicability of classical pattern recognition methods to the problem in hand. The main part of the chapter is devoted to the theory and applications of Neural Classifers which, together with other learning devices labeled herein as boundary techniques, show significant advantages over classical statistical approaches. In the chapter it is also presented a general method for training statistical learning devices that could act as a solver surrogate. On the other hand Chapter 5 is entirely devoted to another powerful boundary classification method known as the Support Vector Machine, which have some interesting features that make it attractive for applications in reliability analysis. Some algorithms for its use in structural reliability are introduced and evaluated through practical examples, which include time-invariant reliability problems, a stochastic finite element analysis and a stochastic stability case. Chapter 6 deals with the alternative statistical learning approach of regression estimation of the performance function both with Neural Networks and Support Vector Machines. Finally, Chapter 7 is the third chapter dealing with the classification approach. In this case, the assumption of the knowledge of the joint density function of the basic variables is abandoned and a regard is paid to problems in which the only available statistical information is of second order. In this case use is commonly made of reliability indices. For this reason, the chapter discusses first the most relevant indices reported in the international literature, under the light of some ideal properties. After showing that no single index fulfills them all, the feasibility of defining reliability indices on the basis of the classification regard on the structural reliability problem is studied. Two indices that can be regarded as extensions of the widely applied Hasofer-Lind (or geometric) index are proposed. The first is based on the simple fact that a linear classifier fitted to classes with a nonlinear boundary is necessarily a secant hyperplane, which is evidently preferable to the tangent one posed by FORM techniques and used in the definition of the geometric reliability index. The second is based on a sophisticated prerogative of Support Vector Machines, that produce a hyperplane in a different space from the input one. Such a plane is then used to define a volumetric index which, as well as the secant one, measures the extension of the safe domain and satisfies all the desirable properties.

The first draft of this book was written for a course given at the Institute of Mechanics of the University of Innsbruck, Austria, by a kind invitation of Professor G. I. Schuëller. The research contained in it has been supported along the last years by the National University of Colombia. Parts of it have also been developed in the framework of research projects carried out by the International Center of Numerical Methods in Engineering of Barcelona, Spain, in which the author has been kindly invited to participate. To these persons and institutions the author expresses his acknowledgements. Special thanks are due to Mr. Diego Álvarez and Mr. Ricardo Henao, who wrote some computer codes employed in this investigation and performed the computations required in some examples.

Manizales,
January 2004 *Jorge E. Hurtado*

Contents

1

A Discussion on Structural Reliability Methods

Statistical Learning methods are very useful for building adequate surrogates for the numerical solvers needed in a reliability analysis of a complex structure. This first chapter has the aim of demonstrating the relevance of such an approximation from several points of view. The chapter is also a critical discussion of reliability analysis methods published up to date.

1.1 Performance and Limit State Functions

The main problem in stochastic mechanics is the estimation of the probability density function of one or several structural responses. If the case under analysis is a dynamical one, then the aim would be to calculate the evolution of such densities. This kind of problem can be termed *full probabilistic analysis*. A derived, more practical problem is to find a specific probability of exceeding a response threshold that can be considered critical for the serviceability or the stability of the system. This is the *reliability analysis*. Notice that, in principle at least, the second problem could be solved after the first, as the probability of failure can eventually be obtained by integration of the probability density function of the observed response. However, most reliability methods developed in the last decades attempt to estimate directly the failure probability or related reliability indices without calculating first the response density function for reasons that are explained in the sequel.

In the reliability problem the input random variables (called also *basic variables*) are collected in a set $\underline{\boldsymbol{x}}$, whose deterministic counterpart is the vector $\boldsymbol{x}$[1]. This defines the *input space*. Let us define a *performance function* $g(\boldsymbol{x})$, in such a way that $g(\boldsymbol{x}) > 0$ means that the sample $\boldsymbol{x}$ is in the safe

[1] In this book an underlined letter indicates a random variable – a vector, if it is written boldface. On the other hand, non-underlined letters are used to denote either their realizations, their deterministic counterparts or deterministic variables in general.

domain and $g(\boldsymbol{x}) < 0$ implies failure. Hence there is a boundary $g(\boldsymbol{x}) = 0$ between the two domains $\mathcal{S} = \{\boldsymbol{x} : g(\boldsymbol{x}) > 0\}$ and $\mathcal{F} = \{\boldsymbol{x} : g(\boldsymbol{x}) \leq 0\}$ that is termed *limit-state function.*[2]

For structural components, the performance function writes

$$g(\boldsymbol{x}) = S(\boldsymbol{x}) - R(\boldsymbol{x}) \ , \tag{1.1}$$

where $\boldsymbol{x} = [x_1, x_2, \ldots, x_d]$ is the vector of d input variables, $S(\boldsymbol{x})$ is a load effect on the component and $R(\boldsymbol{x})$ the strength capacity to withstand it. In the case of complex structures the performance function is given only implicitly through a finite element (or similar) code in the form

$$g(\boldsymbol{x}) = \bar{r} - r(\boldsymbol{x}) = 0 \ , \tag{1.2}$$

in which $r(\boldsymbol{x})$ the implicit function and $\bar{r}$ is a critical threshold for the response $r(\cdot)$ whose surpassing indicates failure. Notice, however, that for structural components the load effects $S(\boldsymbol{x})$ in (1.1) depend on the overall structural response and hence they are also known only implicitly. (An exception to this rule are the simple statically determinate structures, as is well known). As a consequence, a correct probabilistic description of the load effects on each component also requires some solutions of the numerical model of the structure in order to determine the density of $\underline{S}(\boldsymbol{x})$. These remarks point out the relevance of of approximating the structural responses by simple but accurate equations that lead to avoid the repeated use of finite element solvers.

For solving the above reliability problem two kinds of methods have been developed, according to which function is used as a reference. In the *methods based on the limit-state function* the shape of the contour $g(\boldsymbol{x}) = 0$ in the $\boldsymbol{x}$−space is of paramount importance, since the purpose is to compute the integral

$$P_{\mathrm{f}} = \int_{g(\boldsymbol{x}) \leq 0} p_{\underline{\boldsymbol{x}}}(\boldsymbol{x}) \mathrm{d}\boldsymbol{x} \tag{1.3}$$

that yields the probability of failure. Here $p_{\underline{\boldsymbol{x}}}(\boldsymbol{x})$ is the joint probability of the random variables. In this family we have the methods which are most widely applied, such as those based on the Taylor expansion of the performance function (FORM and SORM) and some variants of Monte Carlo methods (Importance Sampling, Directional Simulation, etc). In the *methods based on the performance function* one observes the random variable $\underline{g}(\underline{\boldsymbol{x}})$ and defines the failure probability as

$$P_{\mathrm{f}} = P[\underline{g} \leq 0] \ . \tag{1.4}$$

[2]In some structural reliability literature the expression *limit-state function* is indistinctly used for both functions. We prefer to distinguish between them as this is important from the point of view of statistical learning adopted in this book.

Notice that to this purpose it is not necessary to consider the form of the limit-state function $g(\boldsymbol{x}) = 0$ in the $\boldsymbol{x}$ space. In principle, any stochastic mechanics method giving the probability density function of the observed response, normally through moments or cumulants, can be used for estimating the reliability by integration. Although, most reliability methods published to-date belong to the first category. This is perhaps due to the fact that in this approach there are some byproduct measures such as reliability indices and, in particular, design points that provide useful insight into the structural reliability problem, that are not obtained with the performance function approach. This is especially true in the particular case of incomplete statistical information. For this reason the following discussion of reliability methods is biased to this kind of approaches. However, the Statistical Learning techniques that are the subject of this book can be used either as an approach based on the limit-state function (Chaps. 4 and 5) or, more generally, on the performance function (Chap. 6) and therefore some comments on the latter kind of methods are made at the end of this chapter.

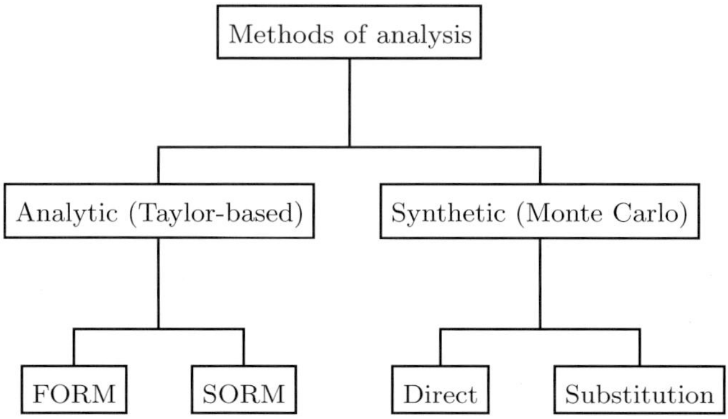

Fig. 1.1. Methods based on the limit-state function

1.2 Methods Based on the Limit State Function

The methods based on the limit-state function that have been proposed so far in the international literature can be grouped into two general families which can be named as *analytic* and *synthetic*, according to whether the random variable set $\underline{\boldsymbol{x}}$ and its effects are treated with the tools of Probability Theory or with those of Statistics (see Fig. 1.1). In the first case we have two families of methods based on the Taylor-series expansion of the limit state function, which are known as FORM (First-Order) and SORM (Second-Order Reliability Method). As is well known, the FORM method requires information on the value of the performance function and its gradient in the vicinity of the so-called *design point* and the SORM method needs in addition the knowledge of the second-order derivatives. In the second case, sample sets of the basic variables are generated and processed by a numerical solver of the structural model in order to obtain a population of structural responses over which statistical analyses can be conducted (Monte Carlo simulation). An important consequence of this approach is that the failure probability becomes a random variable with normally a high coefficient of variation.

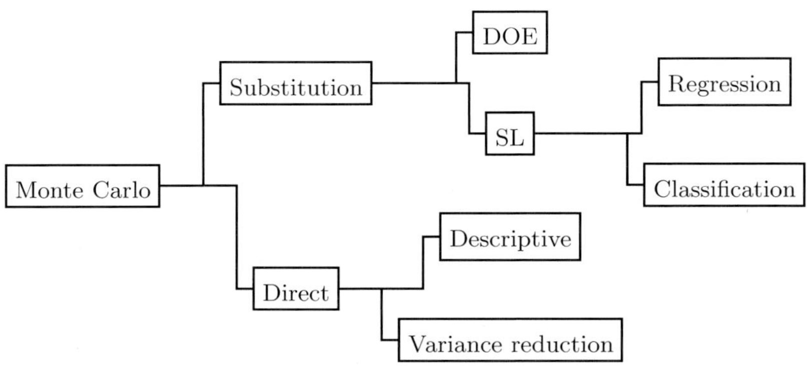

Fig. 1.2. Taxonomy of Monte Carlo methods

Monte Carlo methods in structural reliability (see Fig. 1.2) comprehend those oriented to an overall statistical description of the response, such as Latin Hypercube [86, 49, 76] or Descriptive Sampling [191], and those specially purported at solving the integral (1.3), which are commonly known as *variance reduction methods*. These include Importance Sampling [64, 150] and Directional Sampling [40]. Advanced versions of the descriptive methods, however, have been recently proposed for the reliability problem [122]. All these methods can be applied either to the numerical solver of the structural model to obtain the samples of the structural responses (*direct simulation*) or to a suitable *substitute* thereof (Fig. 1.2). In this latter case a simple function that yields estimates of the structural response of interest in the reliability analysis is calculated. The randomness of the failure probability points out the importance of this alternative because the coefficient of variation of the failure probability tends to be the higher, the lower the probability [9]. If the solver is substituted by an approximating function, the additional runs required to reduce the confidence interval of the estimates can be performed without resort to the finite element code but using the function instead.

The construction of a surrogate of the finite-element solver has been attempted in the last decades by means of the Response Surface Method (RSM) developed in the field of Design of Experiments (DOE) (see e.g. [18, 119]). The main subject of this book is the construction of alternative solver-surrogates using Statistical Learning (SL) techniques such as Neural Networks or Support Vector Machines, which emerge as alternatives to such purpose with several distinguishing advantages. One of them is the possibility of approaching the reliability problem either with regression or with classification tools, as shown later in this chapter.

After introducing these classification of the reliability methods, a summary of their basic traits and limitations will be examined in the next paragraphs.

1.3 Transformation of Basic Variables

The application of some reliability methods, such as FORM and SORM, requires that the basic variable vector $\underline{\boldsymbol{x}}$, defined by its joint density function or an approximate model thereof, be transformed into a set of independent Normal variables $\underline{\boldsymbol{u}}$. Such transformation is also useful when applying Statistical Learning tools such as those studied in the subsequent chapters. This is due to the fact that the application of most statistical learning methods is improved when a preprocess oriented to eliminate the correlation among samples is applied to the training population. In this sense, the classical transformations applied in structural reliability can be regarded as such preprocess. For this reason it is important to examine briefly the most important techniques. The discussion focuses especially the use of these transformations for generating random samples, because most learning procedures use such a kind of training populations (see Sect. 2.12.4). This is in contrast to methods derived from the field of Experimental Design, such as the Response Surface Method, which normally makes use of deterministic sampling procedures.

1.3.1 Normal Variables

The transformation of the basic variables from $\underline{\boldsymbol{x}}$ to $\underline{\boldsymbol{u}}$ can be performed in several ways, according to the available statistical information. For the simple case of independent Normal variables, the transformation is simply the standardization

$$\underline{u}_i = \frac{\underline{x}_i - \mu_{\underline{x}_i}}{\sigma_{\underline{x}_i}}\ , \tag{1.5}$$

where $\mu_{\underline{x}_i}, \sigma_{\underline{x}_i}$ are respectively the mean and standard deviation of the variable $\underline{x}_i$. If the variables are correlated, the set $\underline{\boldsymbol{u}}$ is obtained by a linear transformation of the form

$$\underline{\boldsymbol{u}} = \boldsymbol{T}(\underline{\boldsymbol{x}} - \boldsymbol{\mu}_{\underline{\boldsymbol{x}}})\ , \tag{1.6}$$

where $\boldsymbol{T}$ is a matrix obtained from either the Choleski or the spectral decomposition of the covariance matrix of $\underline{\boldsymbol{x}}$. The latter method is linked to the Principal Component Analysis summarized in Chap. 3, that allows a direct generation of a small subset of important random variables in a transformed space, thus operating an effective dimensionality reduction. Such an important prerogative is not exhibited by the Choleski decomposition.

1.3.2 Normal Translation

In the case where the variables are non-Normal, but are independent, the conventional Normal translation is applicable:

$$\underline{u}_i = \Phi^{-1}\left(P_{\underline{x}_i}(x_i)\right)\ . \tag{1.7}$$

Here $P_{\underline{x}_i}(x_i)$ is the distribution of $\underline{x}_i$ and $\Phi(\cdot)$ is the standard Normal distribution. A similar transformation using a non-standard Normal distribution instead of the standard one is used in the context of FORM algorithms.

Since there are numerous methods for generating random numbers after non-Normal distributions, 1.7 is rarely used in a Monte Carlo context. It is mostly applied in connection to FORM analysis as the so-called normal tail approximation [39] (see also [63]).

1.3.3 Rosenblatt Transformation

If the joint density function of all variables is known the Rosenblatt transformation can be applied [73]. The transformation is based on the fact that the multivariate distribution $P_{\underline{x}_1,\underline{x}_2,\cdots\underline{x}_d}(x_1,x_2,\cdots x_d)$ is equivalent to $P_{\underline{x}_1}(x_1)P_{\underline{x}_2|\underline{x}_1}(x_2|x_1)\cdots P_{\underline{x}_d|\underline{x}_1,\underline{x}_2,\cdots\underline{x}_{d-1}}(x_d|x_1,x_2,\cdots x_{d-1})$. Therefore, the transformation can be performed in the following recursive form:

$$\begin{aligned}
u_1 &= \Phi^{-1}\left(P_{\underline{x}_1}(x_1)\right)\ ,\\
u_2 &= \Phi^{-1}\left(P_{\underline{x}_2|\underline{x}_1}(x_2|x_1)\right)\ ,\\
&\ \vdots\\
u_d &= \Phi^{-1}\left(P_{\underline{x}_d|\underline{x}_1,\underline{x}_2,\cdots\underline{x}_{d-1}}(x_d|x_1,x_2,\cdots x_{d-1})\right)\ ,
\end{aligned} \tag{1.8}$$

where $P_{\underline{x}_i|\underline{x}_j,\underline{x}_k}(x_i|x_j,x_k)$ is the the conditional distribution of x_i given x_j and x_k. These functions can only be obtained in some particular cases. Also, it must be taken into account that the order of inclusion of the variables in the recursion may facilitate or complicate the solution. These difficulties also affect the practical use of the transformation for generating random numbers that obey the multidimensional model $P_{\underline{x}_1,\underline{x}_2,\cdots\underline{x}_d}(x_1,x_2,\cdots x_d)$.

1.3.4 Nataf Transformation

A more common case is that of variables whose joint density function is not known (in fact, there are only few continuous multivariate probability models (see e.g. [99]) so that it is not likely that any of them fits the problem in hand), but the marginals and the correlation structure are known. To such situation the Nataf transformation [120] has been found to be very convenient in stochastic mechanics [104, 130]. It consists in transforming the given variables $\underline{x}_i$ to

$$z_i = \Phi^{-1}\left[P_{\underline{x}_i}(x_i)\right] . \tag{1.9}$$

The set $\underline{\boldsymbol{z}}$ has zero mean, unit standard deviations and correlation matrix

$$\boldsymbol{R}_{\underline{\boldsymbol{z}}} = \begin{pmatrix} 1 & \bar{\rho}_{12} & \cdots & \bar{\rho}_{1,d} \\ \bar{\rho}_{21} & 1 & \cdots & \bar{\rho}_{2,d} \\ \vdots & \vdots & \ddots & \vdots \\ \cdots & \cdots & \cdots & 1 \end{pmatrix} , \tag{1.10}$$

in which the entries are the solution of the integral equation

$$\rho_{ij} = \int_{-\infty}^{\infty}\int_{-\infty}^{\infty}\left(\frac{x_i-\mu_i}{\sigma_i}\right)\left(\frac{x_j-\mu_j}{\sigma_j}\right)\phi_2(z_i,z_j,\bar{\rho}_{ij})\mathrm{d}z_i\mathrm{d}z_j , \tag{1.11}$$

where ϕ_2 is the standard, two-dimensional Normal density function. Finally, a set of independent Normal variables $\boldsymbol{u}$ can be obtained by the Choleski or spectral decomposition of matrix $\boldsymbol{R}_{\underline{\boldsymbol{z}}}$. Empirical formulas for estimating $\bar{\rho}_{ij}$ after ρ_{ij} have been derived by [104]. The empirical equations can be summarized in the following unified form:

$$F_{ij} = \alpha_1 + \alpha_2\eta_j + \alpha_3\eta_j^2 + \alpha_4\rho + \alpha_5\rho^2 + \\ \alpha_6\rho\eta_j + \alpha_7\eta_i + \alpha_8\eta_i^2 + \alpha_9\rho\eta_i + \alpha_{10}\eta_i\eta_j , \tag{1.12}$$

where $\rho = \rho_{ij}$. The values of the coefficients can be found in [104]. The maximum error reported by the authors with respect to the exact values of the ratio is 4.5%. However, most errors are less than one percent.

The pair-wise generation of n random samples according to this model comprehends the following steps:

1. Generate $2n$ standard Normal samples and collect them in vectors $\underline{\boldsymbol{u}}_i, i = 1, 2$.
2. Perform the Choleski decomposition of the correlation matrix $\boldsymbol{R}_{\underline{\boldsymbol{z}}} = \boldsymbol{L}\boldsymbol{L}^{\mathrm{T}}$.
3. Generate the vectors $\boldsymbol{z}_i, i = 1, 2$ as $\boldsymbol{z}_i = \boldsymbol{L}\boldsymbol{u}_i$.
4. Return the coordinates $\underline{x}_{ij} = P_{\underline{x}_i}^{-1}[\Phi(\underline{z}_{ij})],\ i = 1, 2,\ \ j = 1, 2, \ldots, n.$

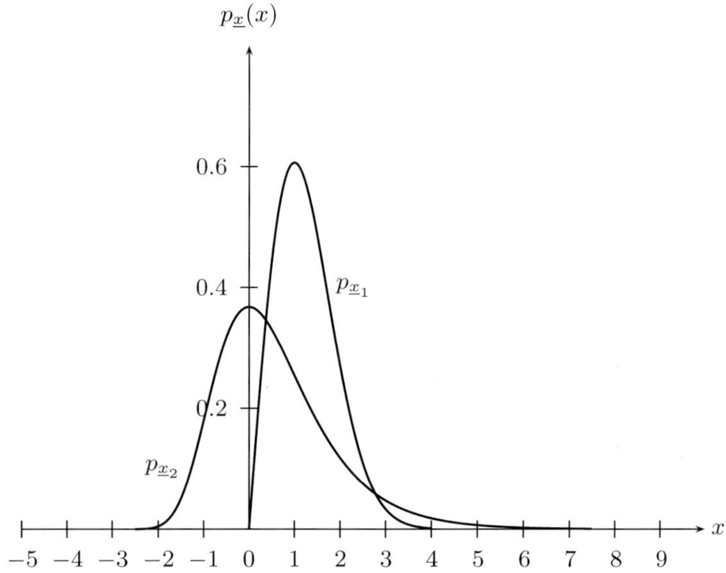

Fig. 1.3. Marginal density functions for building a Nataf model

As an example consider $\underline{x}_1$ and $\underline{x}_2$ be two random variables with a prescribed correlation coefficient $\rho_{12} \equiv \rho = 0.6$, obeying standard Rayleigh and Type I-largest distributions, respectively, as shown in Fig. 1.3. These distributions are

$$P_{\underline{x}_1}(x_1) = 1 - \exp\left(-\frac{x_1^2}{2}\right) ,$$
$$P_{\underline{x}_2}(x_2) = \exp[-\exp(-x_2)] .$$

Accordingly, the density functions are

$$p_{\underline{x}_1}(x_1) = x_1 \exp\Big(-\frac{x_1^2}{2}\Big) ,$$
$$p_{\underline{x}_2}(x_2) = \exp[x_2 - \exp(-x_2)] .$$

The F factor for this case is [104]

$$F = 1.046 - 0.045\rho + 0.006\rho^2 = 1.0212 ,$$

so that the modified correlation coefficient is $\bar{\rho} = 0.6127$. Nataf's estimate of the joint density function is, therefore,

$$p_{\underline{x}_1,\underline{x}_2}(x_1, x_2) = p_{\underline{x}_1}(x_1) p_{\underline{x}_2}(x_2) \frac{\phi_2(x_1, x_2)}{\phi(x_1)\phi(x_2)} ,$$

where

$$\phi_2(x_1, x_2) = \frac{1}{2\pi\sqrt{1-\bar{\rho}^2}} \exp\left[-\frac{x_1^2 - 2\bar{\rho} x_1 x_2 + x_2^2}{2(1-\bar{\rho}^2)}\right] ,$$

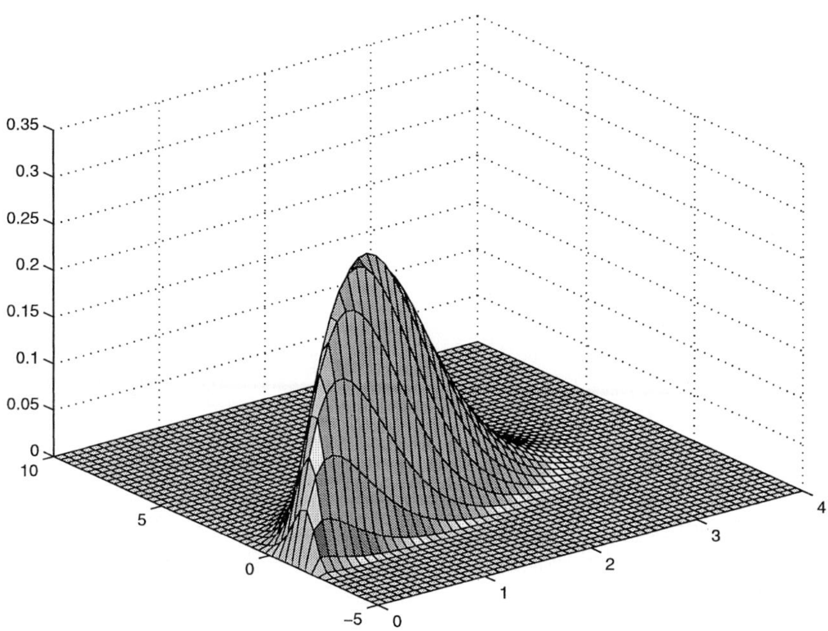

Fig. 1.4. Nataf model for the marginals in Fig. 1.3 with correlation coefficient 0.6

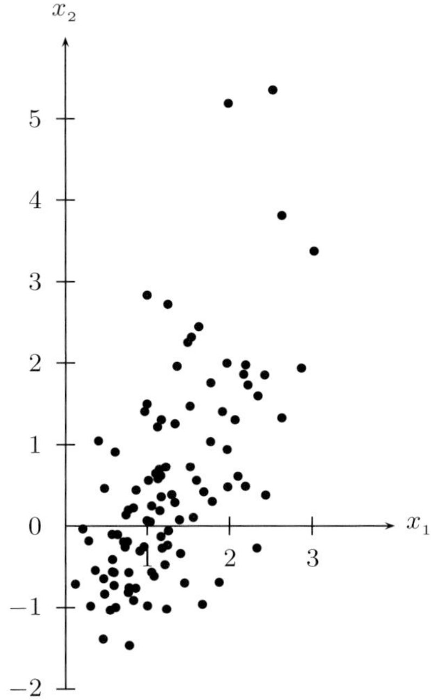

Fig. 1.5. Random samples generated after Nataf model

$$\phi(x_i) = \frac{1}{\sqrt{2\pi}} \exp\left[-\frac{x_i^2}{2}\right], i = 1, 2 ,$$

$$x_1 = P_{\underline{x}_1}^{-1}(\Phi(z_1)) ,$$

$$x_2 = P_{\underline{x}_2}^{-1}(\Phi(z_2)) .$$

Figure 1.4 shows the resulting joint density function. The random variates x_1, x_2 are generated by considering that $\boldsymbol{z} = [z_1 z_2]^{\mathrm{T}}$ is a Gaussian vector with correlation matrix

$$\boldsymbol{R}_{\underline{\boldsymbol{z}}} = \begin{pmatrix} 1 & \bar{\rho} \\ \bar{\rho} & 1 \end{pmatrix} = \boldsymbol{L}\boldsymbol{L}^{\mathrm{T}} .$$

Its Choleski decomposition yields

$$\boldsymbol{L} = \begin{pmatrix} 1 & 0 \\ 0.613 & 0.790 \end{pmatrix} .$$

Hence, realizations $\boldsymbol{z}$ can be generated as

$$\begin{pmatrix} z_1 \\ z_2 \end{pmatrix} = \begin{pmatrix} 1 & 0 \\ 0.613 & 0.790 \end{pmatrix} \begin{pmatrix} u_1 \\ u_2 \end{pmatrix} ,$$

where $\underline{u}_1, \underline{u}_2$ are two independent $N(0,1)$ variates. Finally, the realizations of $\underline{x}_1$ and $\underline{x}_2$ can be computed by calculating the inverse of the respective distribution functions:

$$\begin{aligned} x_1 &= \sqrt{-2\ln(1 - \Phi(z_1)} , \\ x_2 &= -\ln\bigl(-\ln[\Phi(z_2)]\bigr) . \end{aligned}$$

Table 1.1. Exact and empirical statistics for Nataf model example

Statistics	x_1	x_2
Mean	1.271 (**1.253**)	0.504 (**0.577**)
Standard deviation	0.650 (**0.655**)	1.286 (**1.283**)

Figure 1.5 shows a collection of one hundred samples obtained by the above procedure. It can be seen the clear correspondence of this swarm to the model in Fig. 1.4. Specifically, the empirical correlation coefficient is 0.617, which agrees well with the given one (0.6). Other statistics are shown in Table 1.1, in which the figures in bold type are the exact values.

1.3.5 Polynomial Chaoses

A general form of representing a random variable of an arbitrary distribution in terms of Gaussian variables is the polynomial chaos expansion [90]. Generally speaking, a random variable can be expressed as a linear combination of multi-dimensional Hermite polynomials (called in this field polynomial chaoses) having standard Gaussian variables as their argument. In structural reliability the expansion has been applied mainly using one-dimensional Hermite polynomials [40, 183]. In recent years, the technique has become a matter of intensive research in stochastic mechanics as a whole [55, 56, 54, 53].

Formally, a random variable $\underline{x}$ with finite variance admits the representation

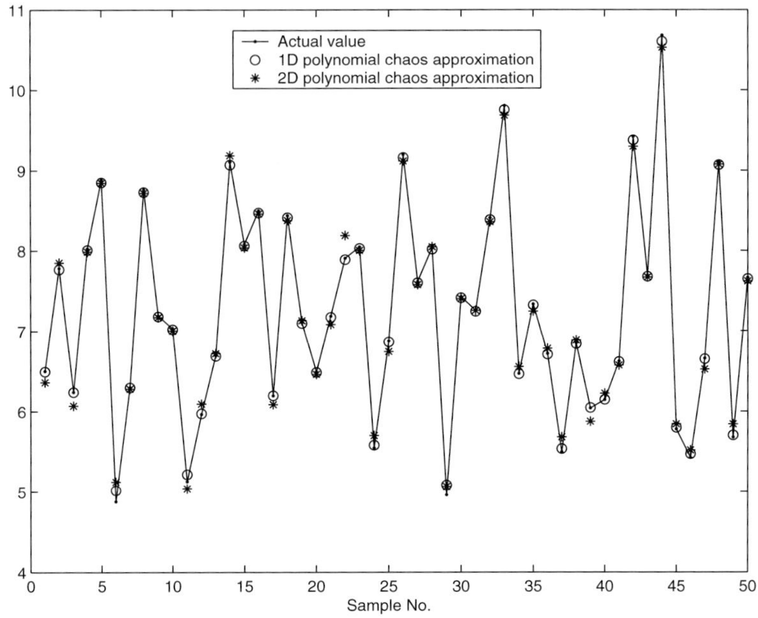

Fig. 1.6. Polynomial chaos approximation of lognormal variates. Sample-by-sample comparison

$$\underline{x} = a_0 \Gamma_0 + \sum_{i_1=1}^{\infty} a_{i_1} \Gamma_1(\underline{u}_{i_1}) + \sum_{i_1=1}^{\infty} \sum_{i_2=1}^{\infty} a_{i_1 i_2} \Gamma_2(\underline{u}_{i_1}, \underline{u}_{i_2}) + \dots , \tag{1.13}$$

where the a's are coefficients, $\underline{u}_{i_j}$ are independent standard Normal variables and $\Gamma_p(\underline{u}_{i_1}, \dots, \underline{u}_{i_r})$ are multidimensional Hermite polynomials, or polynomial chaoses, employed in the analysis of nonlinear systems under the name of Q polynomials [147]. The expansion is convergent in the mean square sense. The polynomial chaoses can be easily obtained by multiplying first-order Hermite polynomials $H_1(\underline{u}_{i_j})$ of each variable $\underline{u}_{i_j}$, except for those variables appearing more than once as argument of the polynomial chaoses, in which case the order of Hermite polynomial is the number of such repetitions. For instance, $\Gamma_3(\underline{u}_1, \underline{u}_1, \underline{u}_2) = H_2(\underline{u}_1) H_1(\underline{u}_2)$. In practice the expansion is applied by truncating the infinite sums. The application of this approach requires the calculation of the coefficients of the expansion, given by

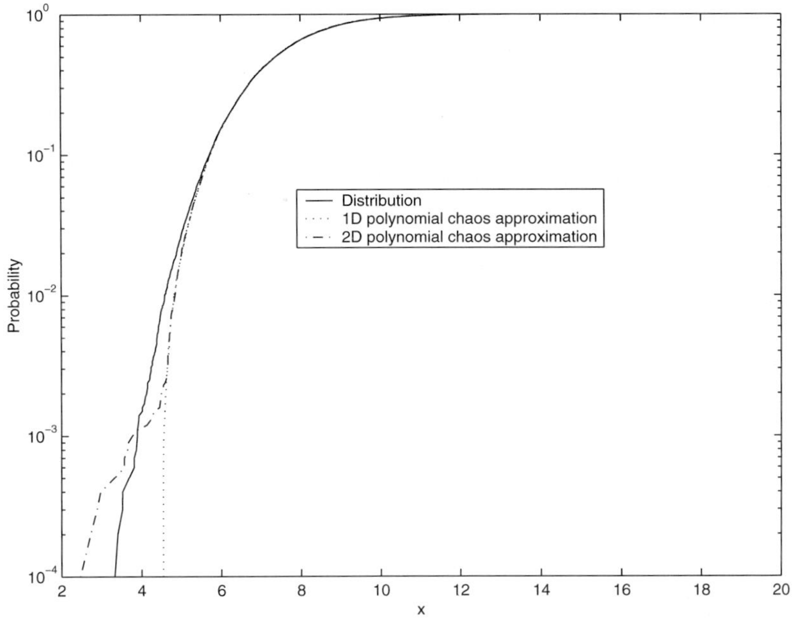

Fig. 1.7. Polynomial chaos approximation of lognormal variates. Comparison of the empirical distributions

$$a_{i_1 i_2 \ldots} = \frac{\mathrm{E}[\underline{x}_j \Gamma_k(\underline{u}_{i_1}, \underline{u}_{i_2}, \ldots)]}{\mathrm{E}[\Gamma_k^2(\underline{u}_{i_1}, \underline{u}_{i_2}, \ldots)]} \,. \tag{1.14}$$

Whereas the denominator of the above expression can be easily obtained using formulas for expectation of products of Gaussian variables, the evaluation of the numerator may be carried out by direct integration or using random numbers. See [147] and [55] for a detailed exposition. Note, however, that the expansion applies only to single random variables. Thus its practical application in structural reliability is hindered by the existence of correlations among the basic variables. In that case the coefficients should satisfy the constraints posed by the cross product expectation of the variables. The so-called Winterstein expansion [183], that has been applied in fatigue analysis, corresponds to the expansion in terms of one-dimensional Hermite polynomials with weights calculated upon such constraints.

Notice that the use of more than one Gaussian variable $\underline{u}_i^{(j)}$ for each basic variable $\boldsymbol{x}_j$ implies an increase of the dimensionality of the problem, which for applying the classification approach to reliability renders positive effects (see

Sects. 2.10 and 5.3). In spite of this and of the good results obtained with the expansion for stochastic finite element analysis, it as such is not very useful as a means for using Gaussian samples $\underline{u}_i^{(j)}$ that represent a variable $\underline{x}_j$, because the convergence of the sum (1.13) is low, especially for extreme values of the variable. In order to illustrate this consider the following example: Ten thousand samples of a lognormal random variable $\underline{x} = \exp(2 + 0.2\underline{u})$ were generated, where $\underline{u}$ is a standard Gaussian variable. Then the coefficients of the expansion up to the fourth order were calculated for the one- and two-dimensional cases using the statistical approximations to the expectations in (1.14). To this end the samples of the Gaussian variable defining the first dimension were calculated as $u_1 = \Phi^{-1}(P_{\underline{x}}(x))$, where $P_{\underline{x}}(x)$ stands for the lognormal distribution and $\Phi(\cdot)$ for the standard Normal distribution. This is because otherwise the convergence would be rather slow. The results for fifty samples appear in Fig. 1.6. It can be seen that there is an improvement in accuracy in passing from the one- to the two-dimensional representation. However, the minor differences affect seriously the distribution of the approximation, as evidenced in Fig. 1.7, especially at the tails. Since extreme values of the basic variables are crucial for the assessment of structural failure probabilities, it is evident that these variables are better represented in a standard Gaussian space by samples obtained with other transformations such as Nataf's. See [60] and especially [46] for a detailed discussion on the accuracy of this representation.

1.4 FORM and SORM

The First- and Second-Order Reliability Methods (FORM and SORM) consist in finding the design point $\boldsymbol{u}^*$ in a suitable standard space $\boldsymbol{u}$ and substituting the actual performance function by its first- or second-order Taylor expansion around that point. In this section a discussion is made about these approaches.

1.4.1 Basic Equations

Notice that the transformation from $\boldsymbol{x}$ to $\boldsymbol{u}$ implies that a performance function $g(\boldsymbol{u})$ substitutes $g(\boldsymbol{x})$[3]. Its second-order Taylor expansion about a point $\boldsymbol{u}^+$ is

$$g(\boldsymbol{u}) \doteq g(\boldsymbol{u}^+) + \boldsymbol{a}^{\mathrm{T}}(\boldsymbol{u} - \boldsymbol{u}^+) + \frac{1}{2}(\boldsymbol{u} - \boldsymbol{u}^+)^{\mathrm{T}}\boldsymbol{H}(\boldsymbol{u} - \boldsymbol{u}^+) , \tag{1.15}$$

where $\boldsymbol{a}$ is the gradient vector evaluated at the expansion point

---

[3]Rigorously, different notations should be used for these two functions, as is done in some papers and books. However, the same symbol $g(\cdot)$ will be used herein for simplicity.

$$\boldsymbol{a} = \nabla g\Big|_{\boldsymbol{u}^+} , \tag{1.16}$$

and $\boldsymbol{H}$ the Hessian matrix

$$\boldsymbol{H}_{ij} = \frac{\partial^2 g}{\partial u_i \partial u_j}\Bigg|_{\boldsymbol{u}^+} . \tag{1.17}$$

In FORM the last term in (1.15) is disregarded and the design point $\boldsymbol{u}^*$ is the solution of the optimization problem

$$\begin{aligned} &\text{minimize } \beta = \left(\boldsymbol{u}^{\mathrm{T}}\boldsymbol{u}\right)^{\frac{1}{2}} \\ &\text{subject to} \quad g(\boldsymbol{u}) = 0 \, . \end{aligned} \tag{1.18}$$

The solution of this optimization problem is carried out by mean of iterative techniques [133]. This problem corresponds to finding the shortest distance from the origin to the limit-state function in the transformed space $\boldsymbol{u}$ such that $g(\boldsymbol{u}^\star) = 0$. The result is

$$\beta = \boldsymbol{w}^{\mathrm{T}}\boldsymbol{u}^\star , \tag{1.19}$$

where

$$\boldsymbol{w} = \frac{\boldsymbol{a}^\star}{|\boldsymbol{a}^\star|} = \frac{\nabla g(\boldsymbol{u}^\star)}{|\nabla g(\boldsymbol{u}^\star)|} \tag{1.20}$$

is the vector normal to the tangent hyperplane at the design point. Hence the equation of the hyperplane becomes

$$\begin{aligned} g(\boldsymbol{u}) &= \boldsymbol{w}^{\mathrm{T}}(\boldsymbol{u} - \boldsymbol{u}^\star) \\ &= \boldsymbol{w}^{\mathrm{T}}\boldsymbol{u} - \beta , \end{aligned} \tag{1.21}$$

which, for comparison with some Statistical Learning equations presented in subsequent chapters, can be put in the form

$$g(\boldsymbol{u}) = \langle \boldsymbol{w}, \boldsymbol{u} \rangle - \beta , \tag{1.22}$$

where $\langle \cdot, \cdot \rangle$ denotes Euclidean inner product.

SORM techniques are based on the above first-order solution. The approximation of the limit state function (1.15) can be presented in terms of β and $\boldsymbol{u}^\star$ as follows:

$$g(\boldsymbol{u}) \doteq \langle \boldsymbol{w}, \boldsymbol{u} \rangle - \beta + \frac{1}{2}(\boldsymbol{u} - \boldsymbol{u}^\star)^{\mathrm{T}}\boldsymbol{B}(\boldsymbol{u} - \boldsymbol{u}^\star) = 0 \tag{1.23}$$

where

$$\boldsymbol{B}_{ij} = \frac{1}{|\nabla g(\boldsymbol{u}^\star)|} \boldsymbol{H}_{ij}\Big|_{\boldsymbol{u}^\star} . \tag{1.24}$$

Some approximations to (1.23) in terms of principal curvatures have been suggested [35, 188]. They are quoted here as they provide useful insight into the reliability problem. By a rotation of the $\boldsymbol{u}$ space to space $\boldsymbol{v}$ such that the design point is located in v_d the axis, function (1.23) can be recast in the form [35]

$$-(v_d - \beta) + \frac{1}{2} \begin{pmatrix} \boldsymbol{v}_{d-1} \\ v_d - \beta \end{pmatrix}^{\mathrm{T}} \boldsymbol{A} \begin{pmatrix} \boldsymbol{v}_{d-1} \\ v_d - \beta \end{pmatrix} = 0 , \tag{1.25}$$

where $\boldsymbol{v}_{d-1} = [v_1 v_2 \ldots v_{d-1}]^{\mathrm{T}}$ and $\boldsymbol{A} = \boldsymbol{RBR}^{\mathrm{T}}$, in which $\boldsymbol{R}$ is an orthonormal matrix with $-\boldsymbol{w}$ as its dth row. Solving the above equation for v_d and retaining only the first two order terms yields

$$v_d = \beta + \frac{1}{2} \boldsymbol{v}_{d-1}^{\mathrm{T}} \boldsymbol{A}_{d-1} \boldsymbol{v}_{d-1} , \tag{1.26}$$

where $\boldsymbol{A}_{d-1}$ is formed with the first $d-1$ rows and columns of $\boldsymbol{A}$. By solving the eigenvalue problem of matrix $\boldsymbol{A}_{d-1}$, i.e. $\boldsymbol{A}_{d-1}\boldsymbol{u} = \kappa \boldsymbol{u}$ one obtains the approximation

$$u_d = \beta + \frac{1}{2} \sum_{i=1}^{d-1} \kappa_i u_i^2 , \tag{1.27}$$

where the eigenvalues are called principal curvatures. This definition is disputed by [188] who support their approximation in the concept of principal curvature of differential geometry. Their approximation reads

$$u_d = \beta + \frac{1}{2R} \sum_{i=1}^{d-1} u_i^2 , \tag{1.28}$$

where

$$R = \frac{d-1}{K},$$

$$K = \sum_{i=1}^{d} \boldsymbol{B}_{jj} - \boldsymbol{w}^{\mathrm{T}} \boldsymbol{B} \boldsymbol{w} . \tag{1.29}$$

In these equations K is the sum of principal curvatures (according to differential geometry) and R is the average principal curvature radius.

1.4.2 Discussion

The calculation of the FORM hyperplane requires the computation of the gradients of the performance function with respect to all basic variables. This is due to the fact that the problem is formulated as a constrained optimization task. Since the performance function is normally not given in closed-form, the gradient calculation can be performed by finite-difference strategies, which requires a minimum of $d+1$ finite element solver calls per iteration point [63]. Since the number of iterations is of the order of ten, it is evident that the calculation of FORM solutions via finite-difference strategies is limited to low dimensionality problems. The situation is more severe in the case of SORM because it is necessary to calculate the elements of the Hessian matrix, which due to symmetry are $d(d-1)/2+d$. Such a involved computation is needed even for applying the approximate SORM techniques, as (1.27) and (1.28) clearly indicate.

For these reason some approaches have been developed for simplifying this task, especially that of FORM. The main trend is the calculation of the gradients at the finite-element level, by means of the so-called *perturbation approach* (see e.g. [105]). This implies for instance the calculation of the derivatives of the stiffness matrices with respect to the basic variables or transformations thereof. This task, however, is also affected by the increase of the dimensionality, because the derivative of a $N \times N$ matrix with respect to a $d \times 1$ vector is a $Nd \times Nd$ matrix. Of course, the situation is more severe for using SORM with such low-level derivatives. Similar dimensionality explosions also occur in using other analytical approaches such as those based on polynomial chaoses [55, 162]. In addition, special finite element solvers should be developed in these approaches.

Another aspect that should be considered is the accuracy and stability of these methods. The limitations of FORM in this regard were pointed out by [151]. With respect to SORM [118] argues that there are several instances in which there is an infinite number of SORM quadratic approximations, for a given set of design point and principal curvatures, that have associated probabilities running from 0 to 1. In words, the SORM results are unstable. Mitteau supports this statement on a single two-dimensional example but a rigorous derivation of its domain of validity is presently lacking. By the same time of Mitteau's paper, an ample study on both FORM and SORM was published by [187]. According to the authors the applicability of FORM is limited to curvatures

$$|K| \leq \frac{1}{10\beta} \tag{1.30}$$

and that of SORM to

$$-0.1\left((2+0.6\beta)\sqrt{d-1}+3\right) \leq |K| \leq 0.4\left(\sqrt{d-1}+3\beta\right) . \tag{1.31}$$

This means that FORM is restricted to cases having very large curvature radius, irrespective to the number of basic variables, and its applicability reduces with the increase of the reliability as measured by β. On the other hand, SORM shows a different trend. Its applicability increases with the dimensionality and with the first-order reliability. However, a previous paper by the authors [188] shows that SORM approximate equations are adequate only if the principal curvatures have all the same sign. If this condition is not met, as in the generalized approximation

$$u_d = \beta + \frac{1}{2R} \sum_{i=1}^{d-1} a^i u_i^2, \quad a < 0 \,, \tag{1.32}$$

all SORM formulas fail to give the exact results with a large error. This is but a consequence of the rigidity of the SORM model in the sense defined in a next section.

In addition, despite [187] show that SORM has much wider applicability than FORM, they do not discuss the dimensionality scaling of the number of solver calls. In fact, it is important to note that in their study it is assumed that the gradients and curvatures have been calculated and, therefore, that the FORM problem has been solved. Hence, the dimensionality effects on the computational effort needed for rendering explicit the limit-state function, that were mentioned above, should anyhow be considered. As a conclusion, it can be said that besides the restrictions for applying FORM and SORM in terms of accuracy, both methods are mainly applicable for low-dimensional problems.

Given the superiority of SORM over FORM some alternatives to alleviate its computational cost have been devised. They consist basically in approximating the gradients and curvatures by fitting a quadratic polynomial using an experimental plan which in the Response Surface Methodology is called saturated design [35, 189]. These applications of the Response Surface Method is called point-fitting in the context of SORM analysis. It is illustrated in Fig. 1.8 after the first of the quoted references. In this way, the deficiencies of the Response Surface Method (see [61] and Sect. 1.6.3) are incorporated into SORM. However, better approximations can be obtained with some Statistical Learning methods as shown in the sequel. What is important at the moment for the purposes of this book is the following:

> *For an implicit performance function, practical application of SORM techniques, which are by far superior to FORM, requires the construction of an approximation of the function in the vicinity of the design point.*

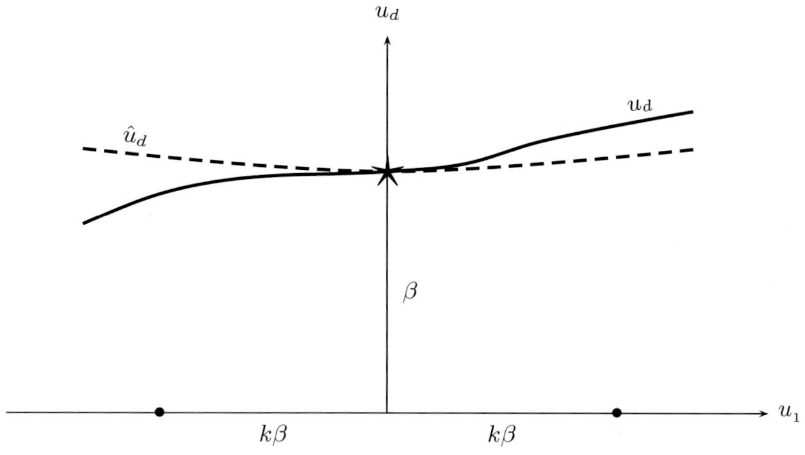

Fig. 1.8. Point-fitting method for SORM approximation (Adapted from [35])

1.5 Monte Carlo Methods

In order to review variance reduction Monte Carlo methods, which are the most commonly applied in structural reliability, let us first summarize the basic (or crude) Monte Carlo simulation technique. For estimating the integral

$$P_{\mathrm{f}} = \int_{\mathcal{F}} p_{\underline{x}}(\boldsymbol{x}) \mathrm{d}\boldsymbol{x} \tag{1.33}$$

the domain of integration can be changed to the entire real space using the indicator function

$$\Upsilon_{\mathcal{F}}(\boldsymbol{x}) = \begin{cases} 1, & \text{if } g(\boldsymbol{x}) \leq 0 \\ 0, & \text{if } g(\boldsymbol{x}) > 0 \, . \end{cases} \tag{1.34}$$

Equation (1.33) then takes the form

$$P_{\mathrm{f}} = \int \Upsilon_{\mathcal{F}}(\boldsymbol{x}) p_{\underline{\boldsymbol{x}}}(\boldsymbol{x}) \mathrm{d}\boldsymbol{x} \, . \tag{1.35}$$

Therefore, the failure probability is the expected value

$$P_{\mathrm{f}} = \mathrm{E}\left(\Upsilon_{\mathcal{F}}(\boldsymbol{x})\right) \, , \tag{1.36}$$

and, consequently, it can be estimated as

$$\hat{P}_{\mathrm{f}} = \frac{1}{n}\sum_{i=1}^{n} \Upsilon_{\mathcal{F}}(\boldsymbol{x}_i) \tag{1.37}$$

using n i.i.d samples $\boldsymbol{x}_i, i = 1, \ldots, n$. This basic Monte Carlo method is known to be excessively costly for reliability analysis because in order to produce one sample in the failure domain it is necessary to obtain a large number of them lying in the safe one. Hence, it is more efficient to modify (1.33) in order to work with a different expected value requiring less samples. Two techniques have been found useful for achieving this goal, namely Importance Sampling and Directional Simulation. In addition, they form the kernel of numerous advanced Monte Carlo methods for structural reliability, some which are mentioned in the next paragraph.

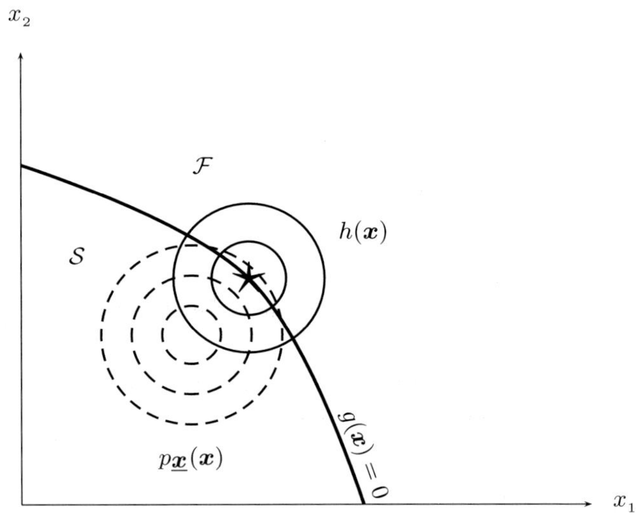

Fig. 1.9. Importance Sampling method

1.5.1 Importance Sampling

In the Importance Sampling (IS) technique (see e.g. [142]) the integral (1.33) is transformed to:

$$P_{\mathrm{f}} = \int_{\mathcal{F}} \frac{p_{\underline{\boldsymbol{x}}}(\boldsymbol{x})}{h(\boldsymbol{x})} h(\boldsymbol{x}) \mathrm{d}\boldsymbol{x} \tag{1.38}$$

or equivalently

$$P_{\mathrm{f}} = \int \varUpsilon_{\mathcal{F}}(\boldsymbol{x}) \frac{p_{\underline{\boldsymbol{x}}}(\boldsymbol{x})}{h(\boldsymbol{x})} h(\boldsymbol{x}) \mathrm{d}\boldsymbol{x} , \tag{1.39}$$

where $h(\boldsymbol{x})$ is an auxiliary density function intended to produce samples in the region that contributes most to the integral (1.33). This transformation implies that the estimate of the failure probability becomes

$$\hat{P}_{\mathrm{f}} = \frac{1}{n} \sum_{i=1}^{n} \varUpsilon_{\mathcal{F}}(\boldsymbol{v}_i) \frac{p_{\underline{\boldsymbol{x}}}(\boldsymbol{v}_i)}{h(\boldsymbol{v}_i)} , \tag{1.40}$$

where the samples $\boldsymbol{v}_i, i = 1, \ldots, n$ are now generated from $h(\boldsymbol{x})$ instead of $p_{\underline{\boldsymbol{x}}}(\boldsymbol{x})$ as before. It is easy to demonstrate that the optimal IS density equals (see e.g. [113]):

$$h(\boldsymbol{x}) = \frac{\varUpsilon_{\mathcal{F}}(\boldsymbol{x}) p_{\underline{\boldsymbol{x}}}(\boldsymbol{x})}{P_{\mathrm{f}}} , \tag{1.41}$$

which in practice is ineffective, as it depends on the target of the calculation, i.e. P_{f}.

At this point, Importance Sampling methods that have been proposed in the literature can be grouped into two categories:

1. *Importance Sampling with design points.* Early applications of the Importance Sampling concept in structural reliability used a density $h(\boldsymbol{x})$ posed over the design point [64, 151, 24] (See Fig. 1.9). The point is approximately identified by a preliminary Monte Carlo simulation with a few samples. A similar procedure was proposed for the case of multiple design points [112, 113]. In these approaches use is normally made of a single multivariate Normal density (for one design point) or of a linear combination of Gaussians (for several points, such as those arising in system safety analysis).
   The use of a design point as a location for a sampling density is, however, not a good choice in some cases. This can be illustrated by having resort to SORM simplified models, which are useful for testing reliability analysis techniques with simple but realistic examples. If the principal curvatures in (1.27) are all negative, an increasing probability mass flows to the failure domain as the number of dimensions increases, so that a density function placed on the design point does not capture the important region as intended. The opposite situation occurs for positive curvatures. This phenomenon is illustrated by Fig. 1.10 in which a standard Normal model is used as the IS density. The same phenomenon happens in case of the cylindrical safe region

$$g(\boldsymbol{u}) = \eta^2 - \sum_{i=1}^{d} u_i^2 = 0 \tag{1.42}$$

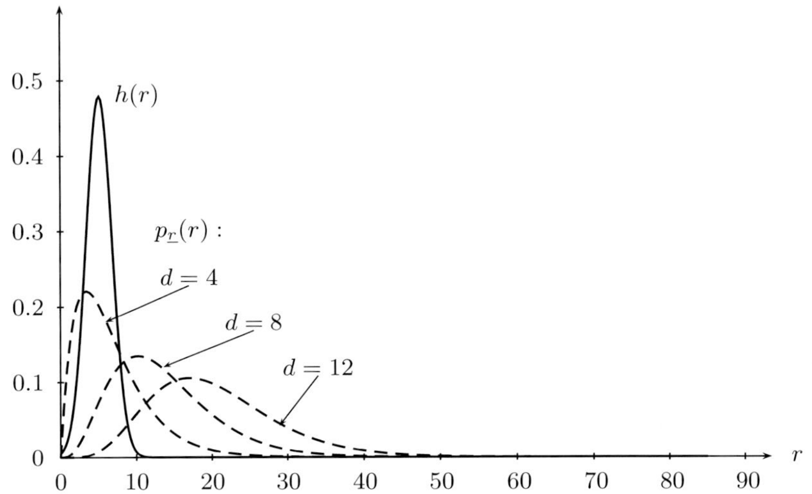

Fig. 1.10. On the inadequacy of Importance Sampling with design points for problems with curvatures of the same sign. For $g(\boldsymbol{u}) = \beta - u_d + \sum u_i^2 = 0$ the variable $r = \sum u_i^2$ is χ^2_{d-1} distributed. An Importance Sampling standard Normal density placed on $u_d = \beta = 3$ does not capture the probability flow of this variable as the dimensionality grows

having the probability of failure

$$P_{\mathrm{f}} = 1 - P[\chi_d^2 \leq \eta^2] \,. \tag{1.43}$$

All this means that in using direct Monte Carlo simulation methods (i.e. without a solver surrogate as indicated by Fig. 1.1) it would be more efficient to locate the *most likely failure point*, defined as that having the highest probability density ordinate in the failure domain, as opposed to the *design point*, defined as that closest to the origin in the standard space.

It can be argued that the Normal density could be made adaptive according to problem dimensionality by increasing the diagonal elements of the covariance matrix. However, according to a recent research by [9], deviations from the standard model are not recommended for a large number of dimensions, since this leads to large deviations of the IS density from its optimal value (1.41). In fact, in the quoted paper the authors conclude that the IS method is applicable with single or multiple design points in the standard $\boldsymbol{u}$ space with many dimensions if the covariance matrix of the Gaussian IS density does not deviate from the identity matrix. However, the authors do not discuss the problem posed by principal curvatures of equal sign, which makes the IS method with design points altogether ineffective as the dimensionality grows.

2. *Adaptive Importance Sampling strategies.* In this group we have several techniques aimed at establishing an IS density as close as possible to the optimal value given by (1.41) [3, 22, 7]. This option is also disputable for large dimensions. In order to show this it is important to compare the estimates of the failure probability given by the basic Monte Carlo method and Importance Sampling (1.37 and 1.40, respectively). Note that while the latter requires the estimation of a density function that approaches the optimal one, the former does not. The problem of generating samples after the optimal IS density without the knowledge of P_{f} can be circumvented by means of simulation algorithms of the Metropolis family (see e.g. [140]), because they operate with a probability ratio in which P_{f} cancels out. This technique has been applied in [7]. However, it remains the problem that an accurate estimation of a density function is seriously affected by the curse of dimensionality, as illustrated by [152]. Let us quote an example conveyed by this author. A direct measure of the accuracy of a density estimate from a given set of samples is the so-called *equivalent sample size* $n_d(\epsilon)$, which is the number of samples in d dimensions necessary to attain an error ϵ in using conventional kernel or finite-mixture density estimates such as those used by [3] and [7]. The crucial point is that the equivalent sample size grows exponentially with problem's dimension. [152] exposes an example in which $n_d(\epsilon) = 50$ in $\mathbb{R}^1$, $n_d(\epsilon) = 258$ in $\mathbb{R}^2$, $n_d(\epsilon) = 1126$ in $\mathbb{R}^3$, etc, using the so-called Epanechnikov criterion for density estimation accuracy. Therefore, if use is made of a low number of samples in (1.40), the errors in estimating $h(\boldsymbol{x})$ will be large. Taking into account that this factor enters in the denominator of the equation, while in the numerator there is a density given beforehand, it is concluded that the introduction of a density estimation problem in the way of calculating a single probability is equivalent to solve a more complicated problem than the target one as an intermediate step. This also occurs in applying some classification methods, as shown in Chap. 4.

1.5.2 Directional Simulation

Directional Simulation is based on the concept of conditional probability. It also exploits the symmetry of the standard space $\boldsymbol{u}$ [17]. Referring to Fig. 1.11, the probability of failure can be expressed as

$$P_{\mathrm{f}} = \int\limits_{\boldsymbol{a} \in \Omega_d} P[g(r\underline{\boldsymbol{a}}) \leq 0 | \underline{\boldsymbol{a}} = \boldsymbol{a}] p_{\underline{\boldsymbol{a}}}(\boldsymbol{a}) \mathrm{d}\boldsymbol{a} \,, \tag{1.44}$$

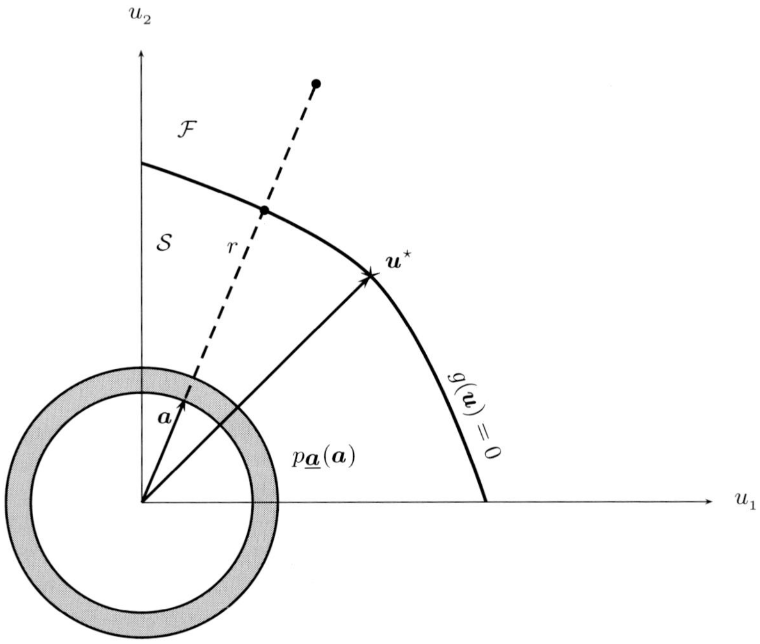

Fig. 1.11. Directional Simulation method

where Ω_d is a unit sphere centered at the origin of the $d-$dimensional space, $\underline{\boldsymbol{a}}$ is a random unit vector in the sphere sampled with a uniform density $p_{\underline{\boldsymbol{a}}}(\boldsymbol{a})$ and r is the solution of $g(r\boldsymbol{a}) = 0$. In this simple directional simulation approach, the probability of failure is estimated as

$$\hat{P}_{\mathrm{f}} = \frac{1}{n} \sum_{i=1}^{n} P[g(r\boldsymbol{a}_i) \leq 0] \tag{1.45}$$

with n samples of $\underline{\boldsymbol{a}}$. If there is only one root r then the probability $P[g(r\boldsymbol{a}_i) \leq 0]$ is equivalent to $P[\underline{r} > r]$ and hence the probability of failure for each sample is given by (1.43). Therefore, (1.45) becomes

$$\hat{P}_{\mathrm{f}} = \frac{1}{n} \sum_{i=1}^{n} \left(1 - P[\chi_d^2 \leq r^2]\right) \ . \tag{1.46}$$

Notice that in this approach the case of curvatures with the same sign poses no difficulties as they are incorporated in the formulation by having resort to the unit sphere. More sophisticated alternatives involve the use of an IS density posed on top of the unit sphere and focused to the design point $\boldsymbol{u}^\star$ [40]. In these alternatives the solution of the equation $g(r\boldsymbol{a}) = 0$, which is difficult in case of implicit performance functions, is substituted by the calculation of a one-dimensional integral. Since the IS density used in them is not placed on the design point but merely oriented towards it, and no optimal IS density is built, the problems of the IS method in high dimensions mentioned above are not present. For this reason, the Directional Simulation method seems theoretically superior than Importance Sampling technique for dealing with high dimensional problems.

1.5.3 General Characteristics of Simulation Methods

After summarizing the most important simulation techniques in the previous paragraphs, it is important to state apart some general features of Monte Carlo methods in order to highlight the relevance of solver-substitution techniques that are briefly examined in the next section and constitute the main subject of this book.

First of all it is important to recall the fundamental advantages of the basic (or crude) Monte Carlo integration method as applied to structural reliability :

- It makes no assumptions about the shape of the limit state function or the location of design points.
- It does not need transformations nor rotations of the basic variable space.
- It is not affected by the so-called curse of dimensionality. In fact, the error rate of crude Monte Carlo integration with n samples is $O(n^{-1/2})$ regardless of the dimensionality of the problem, while other numerical integration methods require $O(n^d)$ samples in order to reach an error rate of $O(n^{-1})$ [103].
- It is entirely general, in the sense that it can be equally applied to linear or nonlinear structural models of any size.
- It does not require the writing of special structural codes as it makes use of standard software employed by the analyst for doing deterministic calculations.

The particular features of advanced Monte Carlo methods have been remarked in the previous paragraphs. However, all Monte Carlo methods as a whole have the disadvantage of being expensive from the computational viewpoint. This is not only due to the fact that it requires the repeated call of the numerical solver for the assessment of any statistical quantity but also that these quantities become random variables. In the particular case of the failure probability, the coefficient of variation of the estimates are, of course, lower when using variance reduction techniques but it tends to be inversely proportional to it. In other words, the lower the failure probability, the higher the spread of the estimates [131, 65, 114, 8]. The coefficient of variation can reach values as high as 0.6 for $P_{\mathrm{f}} = 10^{-4}$, which is a common figure in applications, thus meaning that the one-sigma bounds for it are $[0.4 \times 10^{-4}, 1.6 \times 10^{-4}]$, which is a wide range indeed.

The high cost of Monte Carlo estimates contributes to the refusal of Monte Carlo methods, that is common in structural mechanics fields in which the solution of each deterministic sample takes a large amount of storage and time. In this regard, it is important recall that the advance in the computing capabilities has brought about a parallel increase in the number of degrees of freedom used in finite element calculation and a sophistication of the constitutive models implemented into the structural solvers. Despite the important increase of storage and calculation speed of modern computers that have taken place in recent years, such parallel refining of finite element models has a negative effect in the diffusion on Monte Carlo methods in actual engineering practice.

Since the only way to reduce the confidence interval in Statistics is by increasing the number of samples, after this review of synthetic approaches we arrive to a similar conclusion reached after the discussion of analytical approaches:

> *Practical application of Monte Carlo techniques requires the construction of an approximation of either the performance function or the limit-state function in order to drastically reduce the number of solver calls.*

1.6 Solver Surrogate Methods

Summarizing the conclusions reached after the examination of the development of analytical and synthetic methods, the relevance of building solver surrogates for assisting the reliability analysis of a complex structure is based on the following facts:

- For the reliability analysis at the component or at the structural levels, the limit state function is actually implicit. This hinders the direct application of analytical methods requiring not only the value of the function but also of its gradient and, in the SORM case, the curvatures at the design point. Hence, in spite of FORM and SORM methods are in a sense solver-surrogate techniques, they require much information on the performance function, especially the latter. This information could be provided by a good functional approximation.
- Monte Carlo methods do not require such refined information on the function but need a large number of samples to produce an estimate of the

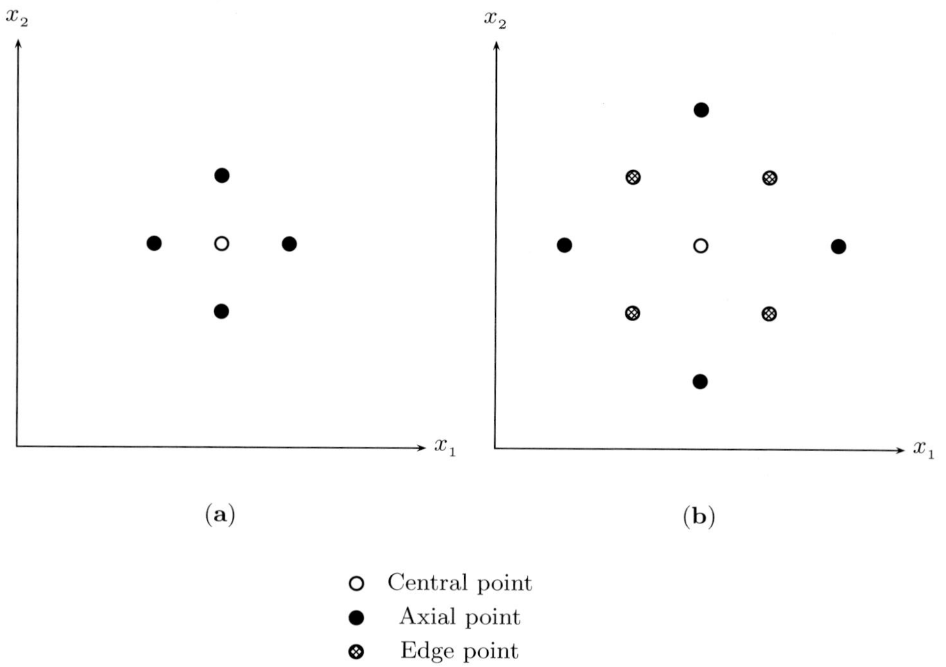

Fig. 1.12. Experimental designs for the Response Surface Method. (**a**): Saturated Design (**b**): Central Composite Design

failure probability. In this respect it must be taken into account that any Monte Carlo estimate is a random variable itself, so that reducing the confidence interval for the mean of the failure probability requires further samples. This task could be drastically reduced if use is made of a good approximation of the implicit functions given by a finite element code.

In addition, it must be said that solver-surrogate methods are not affected by problems of probability flow with dimensionality increase, such as that posed by principal curvatures having the same sign because they are approximations of deterministic functions. For this reason, the design point is in these methods more important than the most likely failure point – in contrast to the Importance Sampling technique.

The task of building solver surrogates has traditionally been pursued by means of the Response Surface Method (RSM) [150, 21, 45, 24, 134], which has been adopted by the structural reliability community after its development in

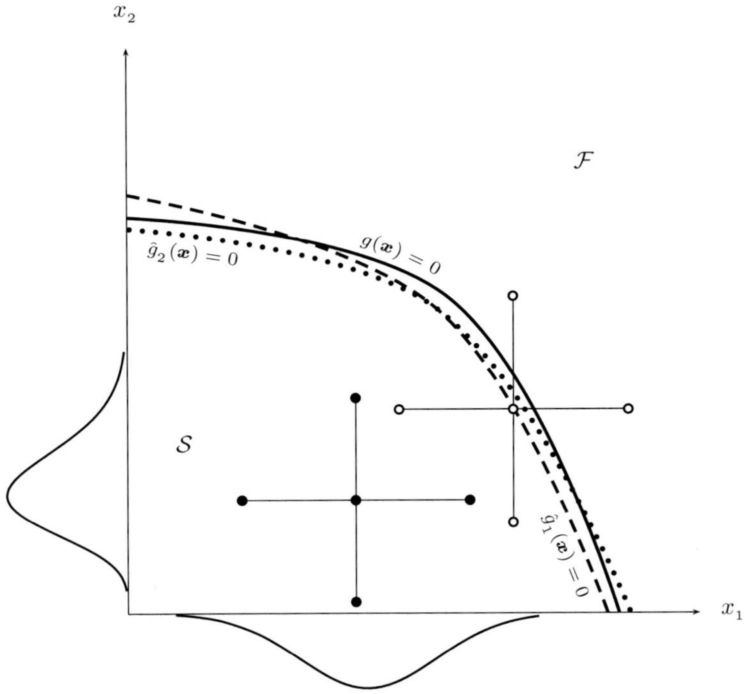

Fig. 1.13. Response Surface Method fitting procedure using Saturated Design. A first experimental plan is placed around the mean vector in the $\boldsymbol{x}$-space. The design point corresponding to the first approximation $\hat{g}_1(\boldsymbol{x}) = 0$ is located, a new plan is placed about it and a new surface $\hat{g}_2(\boldsymbol{x}) = 0$ is calculated

the field of Design of Experiments (see e.g. [119, 18]. Some Statistical Learning techniques, however, can be more accurate and efficient to this purpose. In particular, Neural Networks or Support Vector Machines, which have the characteristic of being flexible and adaptive approaches which are in contrast to the rigidity and non-adaptive nature of the models used in RSM[4].

In the next paragraph the RSM is briefly reviewed. It will be shown that some problems of the method that have been highlighted in recent years [96, 61] are due to the rigid structure of the RSM model and its way of approaching

[4]In this book the expression *Statistical Learning* will be preferred over the more popular and fashionable *Artificial Intelligence*, as this latter points out to biological paradigms and hardware implementations that obscure the purely statistical nature of the in-depth procedures performed by neural networks and other algorithms.

the limit-state function. Next, it is shown why the flexibility and adaptivity of some Statistical Learning techniques make them attractive to overcome these difficulties for the goal of building good approximations of the implicit functions needed in a structural reliability analysis. This description serves as an introduction to the subjects examined at length in the subsequent chapters.

1.6.1 Response Surface Method

The Response Surface Method is purported to replace the structural model solver needed for computing the value of the performance function with a simple expression, which is usually of second-order polynomial form:

$$\hat{g}(\boldsymbol{x}) = \theta_0 + \sum_{i=1}^{d} \theta_i x_i + \sum_{i=1}^{d} \theta_{ii} x_i^2 \tag{1.47}$$

or

$$\hat{g}(\boldsymbol{x}) = \theta_0 + \sum_{i=1}^{d} \theta_i x_i + \sum_{i=1}^{d} \sum_{j=1}^{d} \theta_{ij} x_i x_j \ . \tag{1.48}$$

Here $\boldsymbol{x}$ is the vector of the d input values, $\hat{g}(\cdot)$ is an estimate of the performance function and the θ's are coefficients. The difference between these equations is that the first includes cross terms while the second does not. This function is fitted to a non-random experimental plan by solving a simple algebraic problem. The number of solver calls and the way of finding the coefficients depend on the experimental plan adopted (see Fig. 1.12 and Table 1.2). The coordinates of the experimental vectors in the initial trial are

$$\mu_i \pm k\sigma_i \ , \tag{1.49}$$

where μ_i, σ_i are respectively the mean and standard deviation of the i-th variable and k is a coefficient for the point type. Obviously, $k = 0$ for the central points and typically $k = 1$ for edge points and $k = 2$ to 3 for axial points. After a first trial function, the design point in the standard space can be easily found and successive improvements of the response surface are obtained by locating the center of the experimental plan on the current design point (Fig. 1.13). A total of R runs of the entire program is hence necessary.

The problems found in applying this technique will be reviewed after the next introduction to the Statistical Learning techniques.

1.6.2 Neural Networks and Support Vector Machines

Two classes of Neural Networks have been found to be very useful for building the solver-surrogates [79]: Multi-Layer Perceptrons (MLP) and Radial Basis Function Networks (RBFN). The former implement a function of the form (see e.g. [16])

Table 1.2. Number of solver calls required by the Response Surface Method

Design	Cross terms	Number of calls
Saturated Design	No	$R(2d+1)$
Central Composite Design	Yes	$R(2^d+2d+1)$

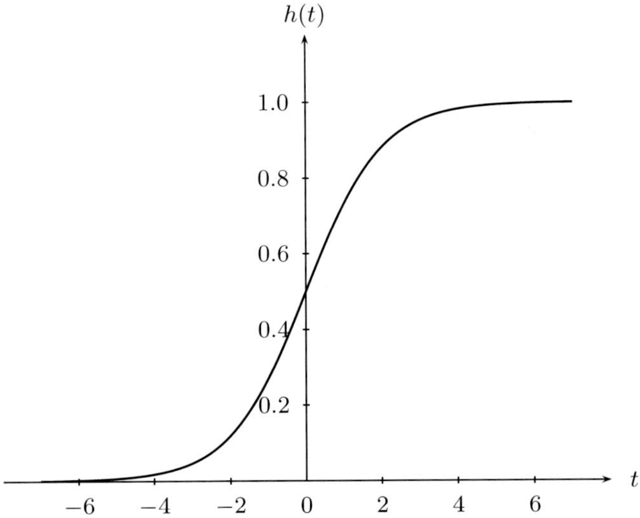

Fig. 1.14. Logistic sigmoid function used by Multi-Layer Perceptrons

$$\hat{g}(\boldsymbol{x}) = \tilde{h}\left(\sum_{k=0}^{m} w_k \; h\left(\sum_{i=0}^{d} w_{ki}x_i\right)\right) , \tag{1.50}$$

where w_k, w_{ki} are weights. The inner sum runs over the number of dimensions d, so that x_i denotes the $i-$th coordinate of vector $\boldsymbol{x}$. On the other hand, $h(\cdot), \tilde{h}(\cdot)$ are nonlinear functions in general. A widely used function h in the inner sum is the logistic sigmoid,

$$h(t) = \frac{1}{1+\exp(-t)} , \tag{1.51}$$

depicted in Fig. 1.14. This function is also used for $\tilde{h}(\cdot)$ in classification analyses while a linear function $\tilde{h}(t) = t$ is commonly applied for regression purposes.

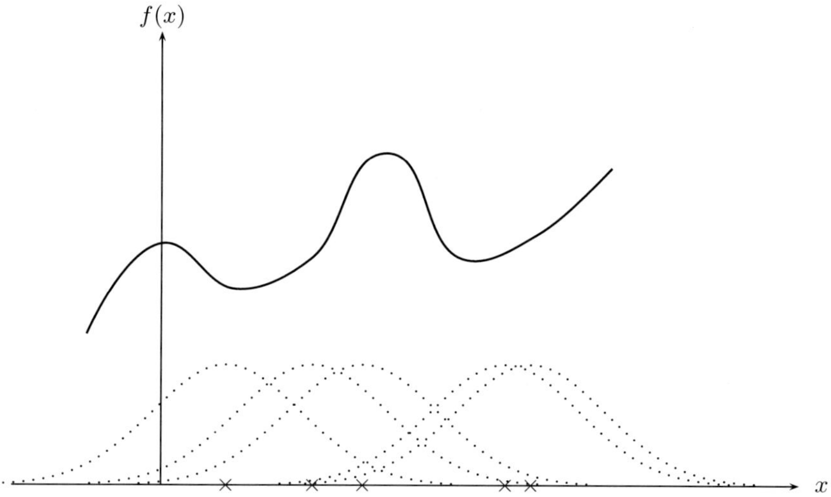

Fig. 1.15. Radial Basis Functions. The crosses indicate the centers at which the radial functions are located

Radial Basis Function Networks are based on a different concept. Instead of working with the coordinates of vector $\boldsymbol{x}$ they use some samples $\boldsymbol{x}_i, i = 1, 2, \ldots, m$ as centers of a symmetrical function in a $d-$dimensional space and build a model of the form

$$\hat{g}(\boldsymbol{x}) = \sum_{i=1}^{m} w_i h(\boldsymbol{x}, \boldsymbol{x}_i) \ , \tag{1.52}$$

where w_i are weights. A common model for the radial function $h(\boldsymbol{x}, \boldsymbol{x}_i)$ is the Gaussian

$$h(\boldsymbol{x}, \boldsymbol{x}_i) = \exp\left(-\frac{||\boldsymbol{x} - \boldsymbol{x}_i||^2}{2\omega^2}\right) \ , \tag{1.53}$$

where the constant ω defines the width of the so-called receptive field of the function. On the other hand, another technique known as Support Vector Machines (SVM) implements a function similar to (1.52) but the weights and number of basis functions are calculated rather differently [174]. The corresponding model is

$$\hat{g}(\boldsymbol{x}) = \sum_{i=1}^{m} w_i K(\boldsymbol{x}, \boldsymbol{x}_i) - b \,, \tag{1.54}$$

where $K(\boldsymbol{x}, \boldsymbol{x}_i)$ is a kernel function and b is a parameter.

The following are the main characteristics of these statistical learning models:

1. *Adaptivity*. By this expression it is meant that the model is composed by a weighted sum of functions whose parameters are determined by the given samples. This is rather evident in the case of RBFN and SVM. In the case of MLP this is achieved indirectly, as the weights which enter as arguments of the functions are determined by the given samples through an optimization algorithm.
2. *Flexibility*. This means that the active region of the basis functions is limited. For instance in the radial basis function (1.53) an input $\boldsymbol{x}$ far from a given center $\boldsymbol{x}_i$ does not cause a meaningful result, whereas other inputs closer to it determine higher outputs. See Fig. 1.15.

The main consequence of these properties considered together is that the model adapts to the given set of samples, so that if these are correctly selected they will have an active participation in the determination of the output. In particular, an important consequence of the adaptivity is that the samples used to fit the model determine both the functions and the weights, thus distributing the effects of incorrect sample selection into a large number of arguments and in a nonlinear form in some of them. On the other hand, the flexibility property implies that a training sample has a limited radius of effect, so that an error in its calculation does not have serious consequences in the entire range of the output.

1.6.3 Characteristics of the Response Surface Method

In contrast to the above SL devices, the RSM works with a model of the form

$$\hat{g}(\boldsymbol{x}) = \sum_{i=1}^{m} w_i h_i(\boldsymbol{x}) \,, \tag{1.55}$$

in which $h_i(\boldsymbol{x})$ are power functions of the coordinates, such as x_j, x_j^2 or $x_j x_k$, with $j, k = 1, 2, \ldots, d$. The RSM model has the following characteristics:

1. *Functions $h_i(\boldsymbol{x})$ are non-adaptive.* In fact, they are not indexed by the given samples of the experimental plan. As a consequence, the effect of this latter is exerted only on the weights w_i, which bear the entire responsibility for the errors. This makes the model highly sensitive to the sample selection as the fitting error is less distributed.
2. *Functions $h_i(\boldsymbol{x})$ are non-flexible.* This is due to the fact that they have infinite active support, in contrast to the radial basis functions, which have a bell shape, or to the sigmoid function, which rapidly saturates.

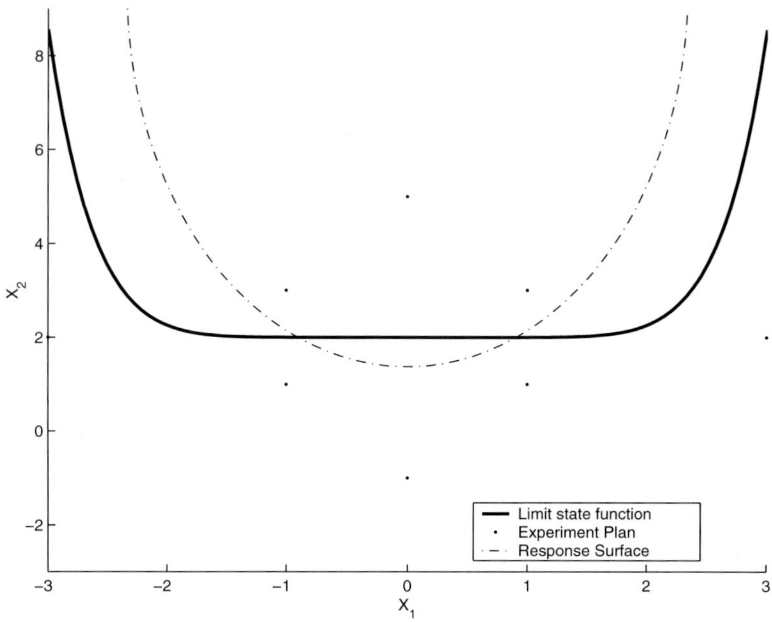

Fig. 1.16. Deficiencies of the RSM for approximating limit state functions. Central composite design with $k = 1$ for the edge points and $k = 3$ for the axial points

In order to illustrate the effects of such rigidity and non-adaptivity, let us consider Figures 1.16 and 1.17, which show two RSM approximations to a nonlinear limit state function

$$g(x_1, x_2) = 4 - 0.001x^8 - 2 \, , \tag{1.56}$$

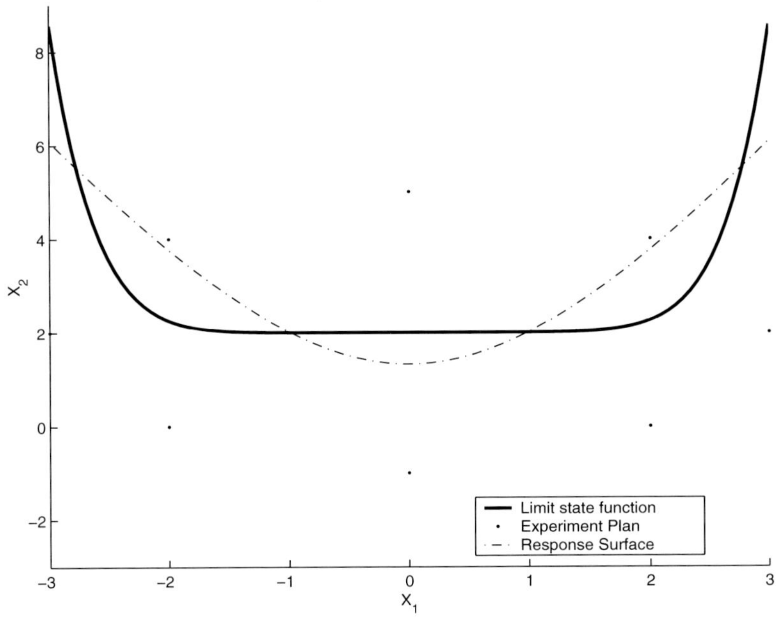

Fig. 1.17. Deficiencies of the RSM for approximating limit state functions. Central composite design with $k = 2$ for the edge points and $k = 3$ for the axial points

discussed by [151] for illustrating the limitations of FORM and by [61] in a paper containing a systematic study on the Response Surface Method. The approximations were obtained by the RSM using the central composite designs shown in the figures and a second order complete polynomial. Notice the important differences in the response surfaces between the two figures. It is evident that the rigidity of the RSM renders difficult the adaptation to an arbitrary, though simple and smooth target function, in spite of the amplitude of the experimental plan. Also, that the resulting surface and hence the estimates of the failure probability and the reliability index are strongly dependent on the experimental plan, as observed in the paper by [61], whose more important conclusions are the following:

- Since the postulated surface depends on the experimental plan (as determined by the parameter k in (1.49), the assessment of the design point (onto which the plan is moved after the first run) will also change. Therefore, the value of the reliability coefficient β and hence the failure probability will also be sensitive to the plan. These values may vary very wildly for small changes of k, as illustrated by Fig. 1.18 for the case of a three bay-five story frame with a dimensionality equal to 21.

- Such wild fluctuations of the reliability measures are also a consequence of the transformation of the basic variables to a standard Normal space required by some reliability methods.

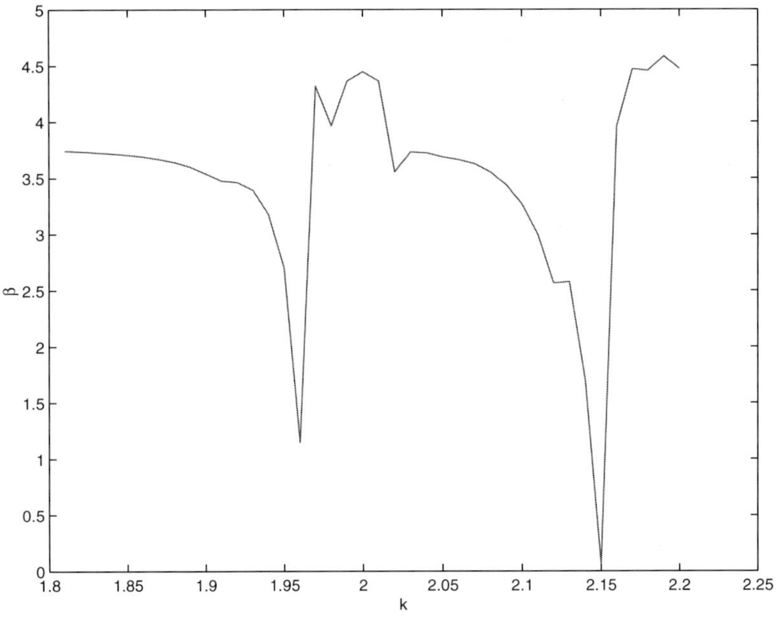

Fig. 1.18. Instability of the RSM estimates of the reliability index β for a three bay-five story frame (After [61])

These problems of the RSM can be attributed to the rigidity of the model. In fact, since the constituent functions are not indexed by the samples, the coefficients of the estimated surface, the trial design point and other quantities are the result of a trade-off (i.e. least square) solution which balances the global errors and assigns all the responsibility to the model weights. Besides, the infinite support of the basis functions causes the fitting errors to affect the entire domain of the basis functions, as is evident from Fig. (1.13). On the contrary, with adaptive-flexible methods using locally active functions the coefficients are balanced with respect to the local errors of the particular function being active in the whereabouts of a certain sample or sample

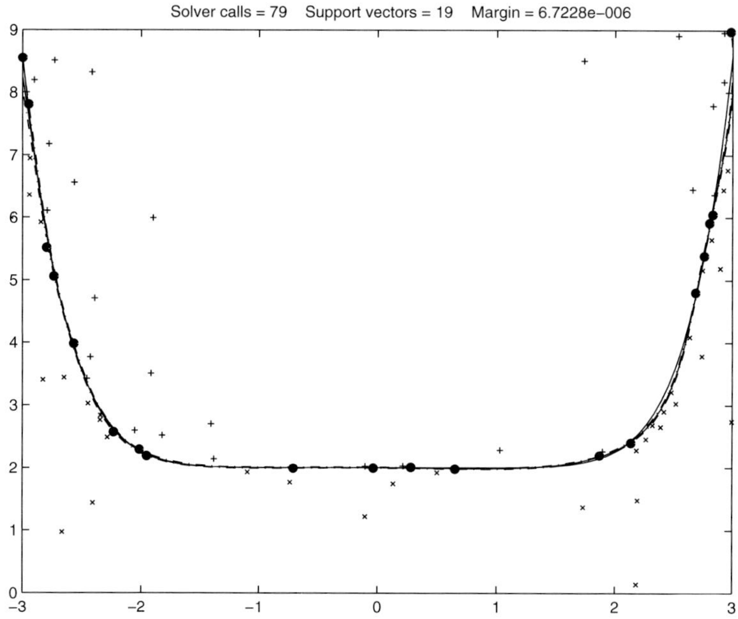

Fig. 1.19. On the advantages of flexible methods for approximating limit state functions. Support Vector approximation of function in Fig. 1.16

group. To illustrate this, Fig. 1.19 shows the approximation to the function (1.56) used above with a support vector machine trained with the algorithm exposed in [80], which uses a random search procedure. Although the SVM approximation required more training samples than the RSM in this case, it is also true that by no means a quadratic polynomial can approximate one of the eighth order with reasonable accuracy. However, a fitting over an ample domain such as that in the figure is necessary when the probability density in the failure region is very flat. For such situations rigid methods like RSM are only adequate if the actual limit state can be well fitted by the imposed model in the entire region; otherwise there is a large risk of error. This discussion illustrates the limitations of rigid models, which at first seem to be more suited for local approximations, i.e. for highly concentrated probability masses such as those arising in problems with highly correlated basic variables. However, the frame example of [61], in which there are some variables highly correlated, shows that there is little hope in that the RSM would be generally applicable even for search domains reduced by high correlation. This is perhaps due to

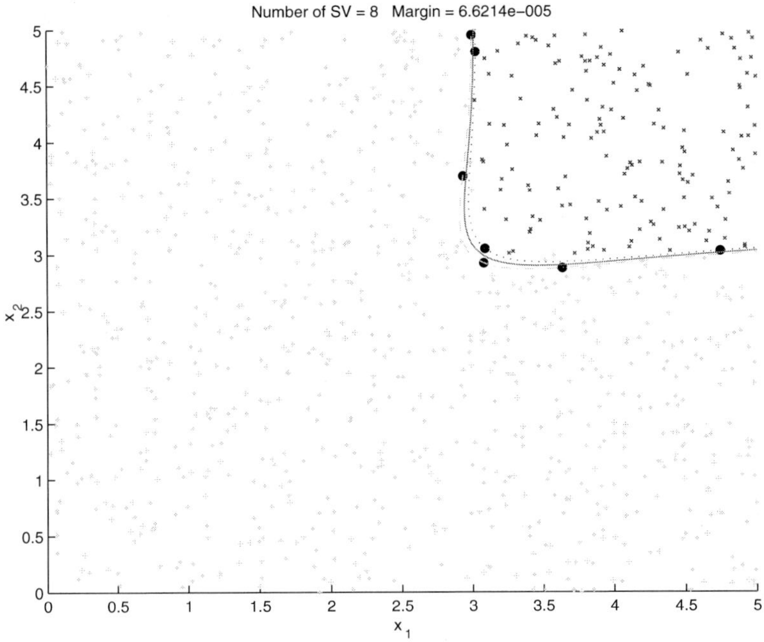

Fig. 1.20. On the advantages of flexible methods for approximating limit state functions. Support Vector approximation for a parallel system problem

the wrong estimation of the trial design points caused by the rigidity of the basis functions of the model.

A situation in which the advantages of flexible over rigid models is evident as well is that of multiple limit state functions, which is common in reliability analysis. In this case the intersections of the functions are often non smooth. Since the rigid RSM model imposes a continuous smooth function, it obviously leads to large errors in the estimation of the failure probability in this case. To cope with that situation, it would be preferable to fit one response surface for each state, a procedure that increases the computational cost. On the contrary, a single approximating function suffices when using flexible models. This is illustrated by Fig. 1.20 which corresponds to the parallel problem $\{\boldsymbol{x} \,|\, \underline{x}_1 > 3 \cap \underline{x}_2 > 3\}$, where $\underline{x}_i, i = 1, 2$ are standard Normal variables. Again, the algorithm reported in [80] for training support vector machines was used. It is observed that the flexible model adapts well to the right angle so that the error in estimating the failure probability in this case is expected to be low.

An analogous contrast between adaptive-flexible models on the one hand and non-adaptive and rigid modes on the other is also exemplified by wavelet and Fourier decompositions of time functions, as shown in the next chapter. In fact, whereas wavelet decomposition is made with basis functions with finite support located about specific time positions, the basis functions used by Fourier analysis have infinite support and have no dependence on specific times.

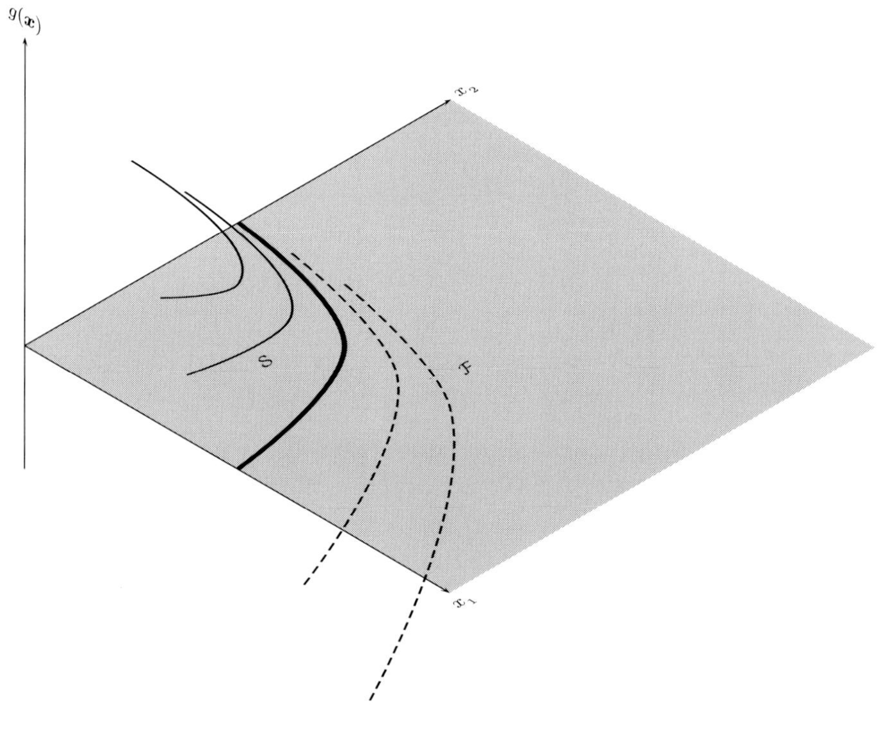

Fig. 1.21. Regression approach for implicit functions

1.7 Regression and Classification

As said in the preceding, building a solver surrogate for the performance function has been traditionally been carried out by means of the RSM. Recently,

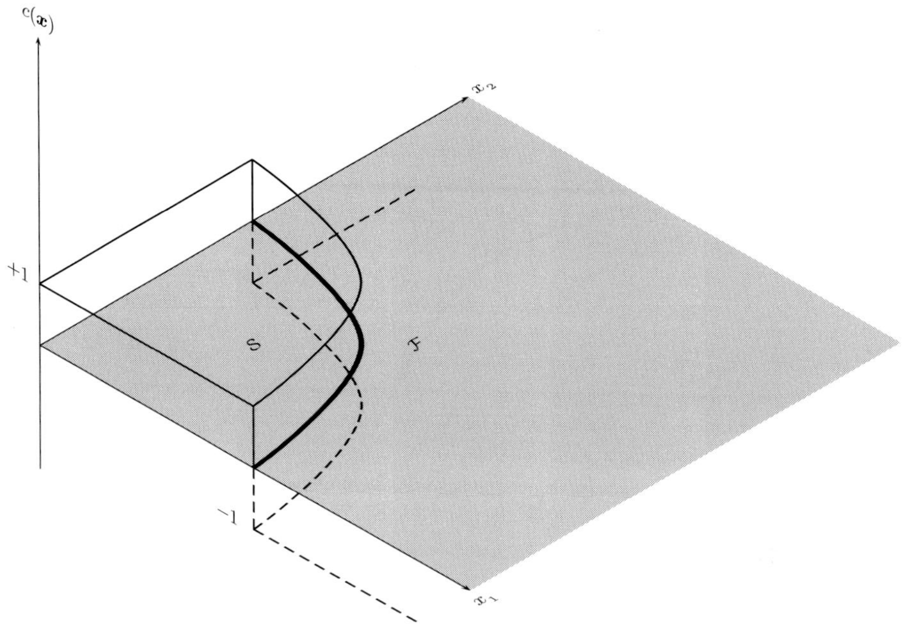

Fig. 1.22. Classification approach for implicit functions

this goal has been pursued by means of Neural Networks, mainly of the MLP type [123, 79, 145, 25]. The author has called the attention to the fact that the radial basis function networks are also useful to this purpose [78, 77].

All these approaches (RSM, MLP and RBFN) are intended to a functional approximation of the performance function, which upon adopting the statistical jargon can be labeled as a *regression approach* (Fig. 1.21), in spite of the fact that this name has a very particular origin with no mathematical meaning at all [153]. However, little attention has been paid by the structural reliability research community to the fact that the problem of rendering explicit the limit-state function and hence estimating the probability of failure can be solved by a *classification approach*, inasmuch as the limit state function separates two well-defined domains, namely the safe and the failure ones (Fig. 1.22). On the basis of some given samples, of which the sign of the performance function is known, the problem becomes that of *pattern recognition*

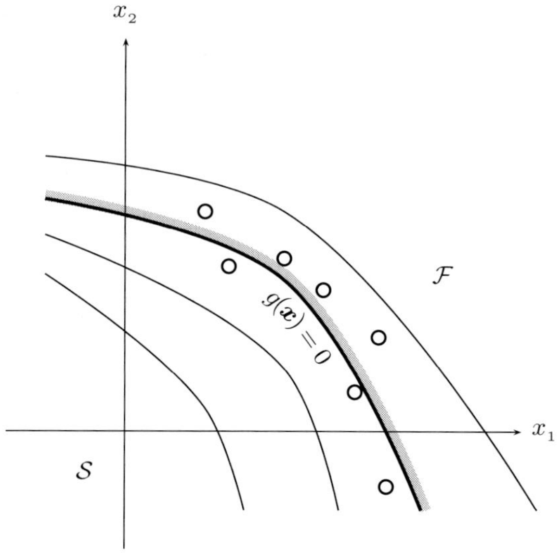

Fig. 1.23. Sampling for limit-state function approximation in the classification approach

(see e.g. [139]), i.e. assigning new incoming samples to one or to the other class.

The classification paradigm has the following advantages:

1. There is no need of requiring high accuracy in the estimation of the performance function values, as only the sign of the function for an specific sample $\boldsymbol{x}$ determines the class to which the sample belongs.
2. The sampling for solver-surrogate building can be concentrated in the vicinity of the limit-state function as this is the actual boundary between the two classes. In other words, while regression approaches require a sampling over an ample domain of the performance function like that shown in Fig. 1.13, classification methods (or at least some of them, as will be shown in Chaps. 4 and 5) are able to construct a classifier with a concentrated sampling like that shown in Fig. 1.23. The geometry of the standard space $\boldsymbol{u}$ can be used to facilitate the generation of such a population as shown later on.

In this regard, it is important to call the reader's attention to the fact that a concentrated sampling such as that shown in Fig. 1.23 was proposed by [96] for building a Response Surface, in a paper that can be considered to be in the direction of overcoming the problems of the RSM by adopting a classification paradigm. Some remarks made in this paper are considered in subsequent chapters.

1.8 FORM and SORM Approximations with Statistical Learning Devices

As shown in the preceding, the FORM and SORM approximations to the limit-state function require information on the derivatives of the performance function at the design point. This information, and especially that required by SORM, is difficult to obtain in case of implicit limit-state functions. However, this task can be facilitated by the approximation of the limit-state function with linear flexible models such as RBFN and SVM that implement models (1.52) and (1.54), which are very similar[5]. Consider for instance the second of them, with the following particular choice of the kernel function (See Chap. 3, Table 3.1 and Chap. 4)

$$K(\boldsymbol{x}, \boldsymbol{y}) = (\langle \boldsymbol{x}, \boldsymbol{y} \rangle + \theta)^p \ , \tag{1.57}$$

where $\langle \cdot, \cdot \rangle$ stands for Euclidean inner product. This is known as the Inhomogeneous Polynomial Kernel in Pattern Recognition literature [148]. Expanding this function for the particular case of $p = 2$ yields

$$K(\boldsymbol{x}, \boldsymbol{y}) = \sum_{n=1}^{d} \sum_{j=1}^{d} x_n x_j y_n y_j + 2\theta \sum_{n=1}^{d} x_n y_n + \theta^2 \ . \tag{1.58}$$

The derivative of the estimate with respect to a coordinate x_k is, therefore,

$$\begin{aligned} \frac{\partial \hat{g}(\boldsymbol{x})}{\partial x_k} &= \sum_{i=1}^{m} w_i \frac{\partial K(\boldsymbol{x}, \boldsymbol{x}_i)}{\partial x_k} \\ &= \sum_{i=1}^{m} w_i \left(\left[\sum_{\substack{n=1 \\ n \neq k}}^{d} x_n x_{kn} x_{in} \right] + 2 x_k x_{ik}^2 + 2\theta x_{ik} \right) \ . \end{aligned} \tag{1.59}$$

In addition,

$$\begin{aligned} \frac{\partial^2 \hat{g}(\boldsymbol{x})}{\partial x_k^2} &= 2 \sum_{i=1}^{m} w_i x_{ik}^2 \ , \\ \frac{\partial^2 \hat{g}(\boldsymbol{x})}{\partial x_k \partial x_j} &= \sum_{i=1}^{m} w_i x_{kj} x_{ij}, \quad j \neq k \ . \end{aligned} \tag{1.60}$$

[5] See though Sect. 5.6.

Similar derivations can be easily carried out for other basis functions. The case of Multi-Layer Perceptrons is, however, more involved, as is evident upon considering (1.50), especially if function $\tilde{h}(\cdot)$ applied at the output layer is nonlinear. However, the Jacobian and Hessian matrices can be easily evaluated by numerical procedures in which use is made of the very network. This is exposed at length in [16].

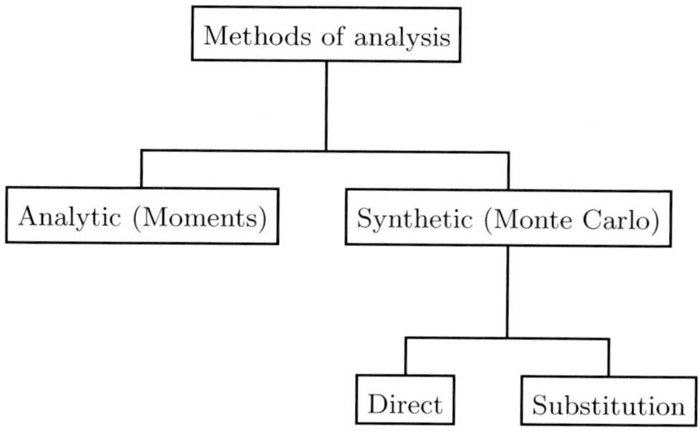

Fig. 1.24. Methods based on the performance function

1.9 Methods Based on the Performance Function

Most methods based on the performance function that have been published thus far are based on the estimation of the density function using moments of the structural response (Fig. 1.24). The moments and/or the density function are obtained either on the basis of the principle of maximum entropy [157, 171, 83], polynomial chaos expansion of the random variables [55], polynomial approximation of the densities [44, 94, 95] or kernel density estimation [5]. Also in this group there are methods aimed at the direct estimation of the reliability with (1.4) via Monte Carlo simulation [74], statistical approximation techniques [67] or Pearson distribution [190] (see, though, [184]).

The Statistical Learning methods briefly introduced in this chapter (MLP, RBFN, SVM) can also be used for this kind of approach acting as solver-surrogates. To this end they are employed as regression functions for substituting the finite-element code in the calculation of the observed responses that make the performance functions. In that case they can provide an estimate of the probability distribution function of such response, as indicated by Table 2.1. Chapter 6 is devoted to a detailed exposition of this approach.

1.10 Summary

In the review of classical reliability methods and their recent developments performed in this chapter, some of their features and shortcomings have been highlighted. The essential conclusions drawn from this review are the following:

- The importance of solver-substitution techniques in structural reliability is supported by the implicit nature of performance functions even at the component level, by the difficulties of calculating the gradients and curvatures in FORM and SORM and by the randomness of the failure probability estimates in Monte Carlo methods.
- The Response Surface Method, which has dominated the research for such a purpose, exhibits some disadvantages when compared to Statistical Learning techniques, such as Neural Networks and Support Vector Machines. In fact, the model used by the method does not have the adaptivity and flexibility of the equations implemented by the SL techniques. The same can be said of the models implemented by FORM and SORM.
- The search for building a substitute to the finite-element solver has been dominated by the functional approximation (or regression) paradigm. However, another way of approaching the problem is by regarding it as a pattern recognition (or classification) task. This paradigm has not been adopted by the structural reliability research community and its exploration is one of the main purposes of the research reported in the present monograph.
- Statistical Learning methods can be used for building regression or classification functions. In the first case they can be used either for obtaining estimates of the probability distribution of the structural responses or the failure probability. In the second case only the latter value can be derived. In other words, the classification paradigm can be used as a method based on the limit-state function for reliability analysis whereas the regression approach is also useful as a method based on the performance function for a full probabilistic analysis.

2

Fundamental Concepts of Statistical Learning

2.1 Introduction

One of the main purposes of applying analytic or synthetic methods in structural reliability is to reduce to a minimum the number of samples, because of their computational cost. This makes relevant the consideration of the Theory of Statistical Learning (SLT) developed in the last decades but specially in the last years[1], as it is aimed to solve, upon the basis of a few available samples, the following problems:

- *Classification*, i.e., the assigning of a sample to classes, which are labeled numerically or symbolically.
- *Regression*, that is, the estimation of the functional dependence of some variables with respect to others.
- *Probability density estimation* of a random variable.

The above problems are ordered according to an increasing order of difficulty. In fact, assigning a sample to a class is (at first sight at least) a simpler task than the estimation of the numerical dependence among the variables composing the population and, in turn, this is clearly easier than the estimation of the probability density of the output variables of a system, which is an ill-posed problem [174].

In this chapter the essential elements of such a theory will be exposed. Under this light, some reliability methods using either classification, regression or density estimation can be examined. The regression approach has been used for reliability analysis as said in the preceding chapter as well as the the density estimation approach, without reference to the framework of SLT, however. First attempts in this direction have been done by the author and co-workers [81, 79, 84, 78], using neural networks as a regression technique.

[1]Important milestones are the works by [141, 176, 101, 75, 12, 57] among many others. A recommended list of books is the following: [16, 139, 174, 175, 4, 27, 42, 69, 148, 70].

The classification approach to reliability analysis in particular, however, has not been followed, with perhaps the only exemption of a first attempt made by the author in [79] using neural classifiers and the proposal of a support vector classification for reliability analysis in [80].

2.2 The Basic Learning Problem

The purpose of statistical learning can be defined as follows: To build a function that performs either the classification, regression or density estimation tasks, upon the basis of some given samples. The function building is known as *learning* or *training* and the functions themselves are sometimes called *machines*, not only because they can be realized as electronic devices, but also for philosophical reasons [175]. In fact, while reflective human reasoning approaches physical problems with strong mathematical models and hypothesis made almost entirely *a priori* (i.e. previous to experience), these machines aim at searching the inner relationships upon almost purely empirical basis.

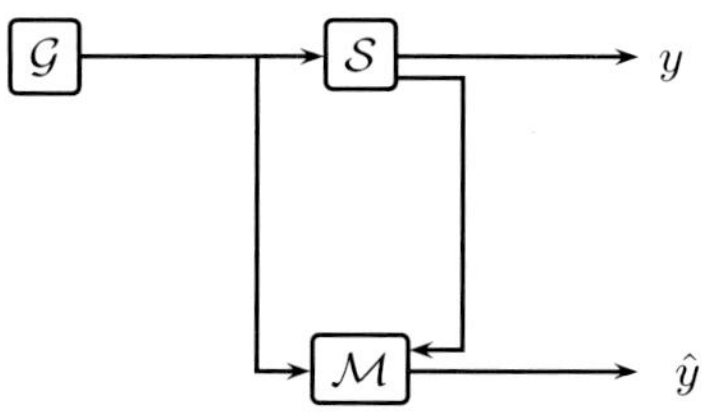

Fig. 2.1. General supervised learning problem (Adapted from [175])

The building of a function on the basis of a few given samples is an instance of the problem of learning from data. This problem can be explained with the aid of the diagram appearing in Figure 2.1. First, there is a generator $\mathcal{G}$ of random vectors $\boldsymbol{x}_i \in \mathbb{R}^d$, which in the reliability analysis is a Monte Carlo simulation algorithm that produces the samples from the joint distribution of the variables $p_{\underline{\boldsymbol{x}}}(\boldsymbol{x})$. There is also a supervisor $\mathcal{S}$ that yields the right label of the class y corresponding to a sample $\boldsymbol{x}$, which obviously corresponds to the finite element solver in structural reliability. Finally, there is a learning machine $\mathcal{M}$, that applies functions $h(\boldsymbol{x}, \boldsymbol{\omega})$ and produces an estimate of the class $\hat{y}$. Here $\boldsymbol{\omega} \in \Omega$ is a vector of parameters that defines the function. There are two distinct problems concerning the learning machine. The first is the correctness of the training, i.e. the error in estimating the results corresponding to the given samples (*training error*); the second is the prediction ability, which concerns the error in assessing the outputs for samples not used in the training phase (*generalization error*). Learning machines such as neural networks, support vector machines and others occupy an intermediate position between parametric and non-parametric models in that they use linear or non-linear combinations of a multiplicity of assigned functions of local activation that are indexed by the samples[2]. With this flexibility they overcome the difficulties which are inherent to parametric models, whose rigidity hinders the adapting to the given data, and to non-parametric ones, which are not well suited for high dimensional spaces.

Fig. 2.2. General unsupervised learning problem

All the above said corresponds to the kind of learning called *supervised* for obvious reasons. To this group there belong techniques for regression and classification such as Multi-Layer Perceptrons [88], Radial Basis Function Networks [185], Support Vector Machines [175] and special types of neural networks used for signal processing and optimization which are beyond the interest of the present monograph [28, 107]. Another kind of learning called *unsupervised* is, however possible (Fig. 2.2). In this case there are no known outputs and hence the purpose of the analysis is to extract a structure $\mathcal{U}$ from the data. To this group belong techniques for kernel density estimation such as the Expectation-Maximization algorithm [34], data compression and dimensionality reduction methods such as Vector Quantization and Self-Organizing Maps [98], Principal Component Analysis [136], Kernel PCA [148], Principal Component Neural Networks [37] and many others.

2.3 Cost and Risk Functions

Let us now define some functions necessary to determine the degree of learning on the basis of samples. The *cost* or *loss* function is defined a function $L(y, h(\boldsymbol{x}, \boldsymbol{\omega}))$ of the estimates obtained by the learning machine and of the

[2]An example of parametric approaches is the maximum likelihood estimation of e.g. the parameters of a Normal density for explaining a given population, whereas a popular non-parametric model is the histogram.

actual results, in such a way that the function grows with the difference between them. On the other hand, the *risk function* is defined as the expected value of the cost function:

$$R(\boldsymbol{\omega}) = \int L(y, h(\boldsymbol{x}, \boldsymbol{\omega}))\mathrm{d}P_{\underline{\boldsymbol{x}},\underline{y}}(\boldsymbol{x}, y) \ , \tag{2.1}$$

where $\mathrm{d}P_{\underline{\boldsymbol{x}},\underline{y}}(\boldsymbol{x}, y)$ is the probability measure of the random variables $\{\boldsymbol{x}, y\}$. If the probability distribution $P_{\underline{\boldsymbol{x}},\underline{y}}(\boldsymbol{x}, y)$ is derivable the above Stieltjes integral reduces to

$$R(\boldsymbol{\omega}) = \int L(y, h(\boldsymbol{x}, \boldsymbol{\omega}))p_{\underline{\boldsymbol{x}},\underline{y}}(\boldsymbol{x}, y)\mathrm{d}\boldsymbol{x}\mathrm{d}y \ , \tag{2.2}$$

where $p_{\underline{\boldsymbol{x}},\underline{y}}(\boldsymbol{x}, y)$ is the density function of the set of variables. Since this latter remains unknown, the above equation can be solved by approximating the actual risk by the empirical estimate $R_\mathrm{e}(\boldsymbol{\omega}, n)$, defined as the average of the cost function over the available samples. Hence it is a function of their quantity. Since in a learning approach on the pure basis of the given samples the use of density models is to be avoided, the selected function $h(\boldsymbol{x}, \boldsymbol{\omega})$ will be that giving the least empirical risk, i.e $\min(R_\mathrm{e}(\boldsymbol{\omega}, n))$. This approximation is known as *Empirical Risk Minimization principle* (ERM).

Equipped with these definitions, we are now in a better position to define the three learning approaches more rigorously:

- *Probability density function estimation* of the set $\underline{\boldsymbol{x}}$. In this case there is no need of considering an actual response y, as it is an unsupervised problem. The cost function is

  $$L(h(\boldsymbol{x}, \boldsymbol{\omega})) = -\ln h(\boldsymbol{x}, \boldsymbol{\omega}) \tag{2.3}$$

  according to the Maximum Likelihood Principle. Hence the risk function takes the form

  $$R(\boldsymbol{\omega}) = -\int \ln h(\boldsymbol{x}, \boldsymbol{\omega})p_{\underline{\boldsymbol{x}}}(\boldsymbol{x})\mathrm{d}\boldsymbol{x} \ . \tag{2.4}$$

  Since at a difference to the classical approach the solution will not be sought on the basis of a density model $p_{\underline{\boldsymbol{x}}}(\boldsymbol{x})$, it is necessary to define the empirical risk using the samples $\boldsymbol{x}_1, \boldsymbol{x}_2, \ldots, \boldsymbol{x}_n$ as

  $$R_\mathrm{e}(\boldsymbol{\omega}, n) = \frac{1}{n}\sum_{i=1}^{n} L(h(\boldsymbol{x}_i, \boldsymbol{\omega})) \ . \tag{2.5}$$

  Note that the ERM principle is in some sense more general than the maximum likelihood principle as it is not tied to a specific density model assumption.

- *Regression.* As is well known, the regression model for functional approximation is the function $g(\boldsymbol{x})$ such that in the identity

$$y = g(\boldsymbol{x}) + \epsilon \tag{2.6}$$

the square of the error ϵ is a minimum. Rigorously, this is equivalent to define the regression function as the expected value

$$g(\boldsymbol{x}) = \int y \; \mathrm{d}P_{\underline{y}|\underline{\boldsymbol{x}}}(y|\boldsymbol{x}) \; . \tag{2.7}$$

For solving the problem we have set of functions $h(\boldsymbol{x}, \boldsymbol{\omega})$ that may or may not contain the target function $g(\boldsymbol{x})$. A value $\boldsymbol{\omega} \in \Omega$ is sought such that it renders the minimum of the cost function

$$L(y, h(\boldsymbol{x}, \boldsymbol{\omega})) = (y - h(\boldsymbol{x}, \boldsymbol{\omega}))^2 \; . \tag{2.8}$$

Hence, the risk function (2.2) takes the form

$$R(\boldsymbol{\omega}) = \int (y - h(\boldsymbol{x}, \boldsymbol{\omega}))^2 \, p_{\underline{\boldsymbol{x}},\underline{y}}(\boldsymbol{x}, y)\mathrm{d}\boldsymbol{x}\mathrm{d}y \tag{2.9}$$

if the density exists. It can be demonstrated [16] that this function can be written in the following form:

$$R(\boldsymbol{\omega}) = \int \mathrm{E}\left(\epsilon^2|\boldsymbol{x}\right) p_{\underline{\boldsymbol{x}}}(\boldsymbol{x})\mathrm{d}\boldsymbol{x} + \int (g(\boldsymbol{x}) - h(\boldsymbol{x}, \boldsymbol{\omega}))^2 \, p_{\underline{\boldsymbol{x}}}(\boldsymbol{x})\mathrm{d}\boldsymbol{x} \; , \tag{2.10}$$

where it has been assumed that the error ϵ has zero mean. In this equation the first term is determined by the variance of the error while the second concerns the accuracy of the estimation. Since the first term does not depend on $\boldsymbol{\omega}$, the minimization reduces to finding the parameters that make the second term a minimum, which occurs when the optimal value of $h(\boldsymbol{x}, \boldsymbol{\omega})$, namely $h(\boldsymbol{x}, \boldsymbol{\omega}^*)$, equals $g(\boldsymbol{x})$.
Let us consider a set of samples $\boldsymbol{q} = \{(\boldsymbol{x}_1, y_1), (\boldsymbol{x}_2, y_2), \ldots, (\boldsymbol{x}_n, y_n)\}$ and present (2.6) in the form

$$y_i = h(\boldsymbol{x}_i, \boldsymbol{\omega}^*) + \epsilon_i \; . \tag{2.11}$$

The likelihood function is then

$$\begin{aligned} V(\boldsymbol{q}|\boldsymbol{\omega}) &= \sum_{i=1}^{n} \ln p_{\underline{\epsilon}}(\epsilon_i) \\ &= \sum_{i=1}^{n} \ln p_{\underline{\epsilon}}(y_i - h(\boldsymbol{x}_i, \boldsymbol{\omega}) \; . \end{aligned} \tag{2.12}$$

If the error is assumed to be Gaussian with zero mean and variance $\sigma_{\underline{\epsilon}}^2$, then

$$V(\boldsymbol{q}|\boldsymbol{\omega}) = -\frac{1}{2\sigma^2}\sum_{i=1}^{n}(y_i - h(\boldsymbol{x}_i,\boldsymbol{\omega}))^2 - n\ln(\sqrt{2\pi}\sigma_{\underline{\epsilon}}) \; . \tag{2.13}$$

Consequently, the maximization of the likelihood is equivalent in this case to the minimization of the empirical risk

$$R_{\mathrm{e}}(\boldsymbol{\omega}, n) = \frac{1}{n}\sum_{i=1}^{n}(y_i - h(\boldsymbol{x}_i,\boldsymbol{\omega}))^2 \; . \tag{2.14}$$

Despite in practice this approach is correct only in the case of Gaussian errors, it is normally used for other (normally unknown) distributions.

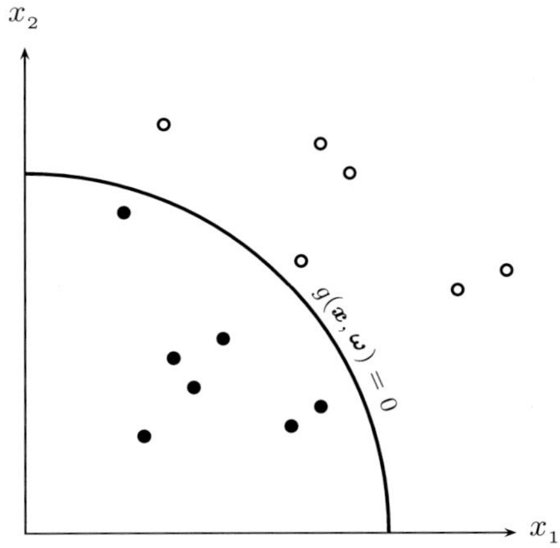

Fig. 2.3. Classification problem

- *Classification.* Let $c(\boldsymbol{x}) = \{+1, -1\}$ be a function that takes values +1 or -1 depending on whether the sample $\boldsymbol{x}$ belongs to one class or to another. A classification machine is sought such that it uses a function $h(\boldsymbol{x},\boldsymbol{\omega})$ that also takes values $\{+1, -1\}$. Hence $h(\boldsymbol{x},\boldsymbol{\omega})$ equals sgn $(g(\boldsymbol{x},\boldsymbol{\omega}))$, where $g(\boldsymbol{x},\boldsymbol{\omega})$ is a function that severs the plane $\boldsymbol{x}$ thus separating the two classes (see Fig. 2.3). The purpose of the learning algorithm is to find the value of $\boldsymbol{\omega}$ that minimizes the classification error computed over the samples. To such purpose let us define the following cost function:

$$L(c, h(\boldsymbol{x},\boldsymbol{\omega})) = \begin{cases} 0, & \text{if } c = h(\boldsymbol{x},\boldsymbol{\omega}) \\ 1, & \text{if } c \neq h(\boldsymbol{x},\boldsymbol{\omega}) \; . \end{cases} \tag{2.15}$$

In this way the error grows as far as there are more divergences between the estimations given by the machine and the actual classes said by the supervisor. The risk function is given by (2.2). Given some samples $(\boldsymbol{x}_1, c_1), (\boldsymbol{x}_2, c_2), \ldots, (\boldsymbol{x}_n, c_n)$, the empirical risk assumes the form

$$R_{\mathrm{e}}(\boldsymbol{\omega}, n) = \frac{1}{n} \sum_{i=1}^{n} \Upsilon\left(c_i = h(\boldsymbol{x}_i, \boldsymbol{\omega})\right) , \tag{2.16}$$

where $\Upsilon(A)$ is the indicator function of a logical statement A:

$$\Upsilon(A) = \begin{cases} 1, & \text{if } A \text{ holds} \\ 0, & \text{otherwise} . \end{cases} \tag{2.17}$$

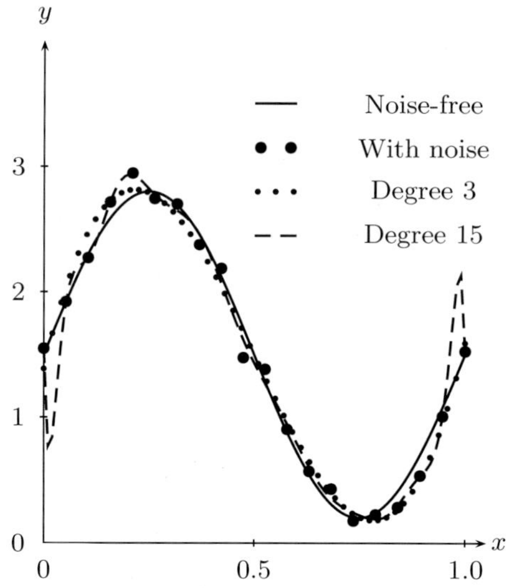

Fig. 2.4. Polynomial regression

2.4 The Regularization Principle

The ERM leads to finding a model that allows estimating the values of a function in the regression problem or the class corresponding to a sample in the classification one. However, in every problem there is always some a priori knowledge which can be exploited for the sake of model improvement.

This a priori knowledge is normally codified in the selection of the form of the approximating functions $h(\boldsymbol{x}, \boldsymbol{\omega}), \boldsymbol{\omega} \in \Omega$. However, the structure of these functions must be subject to control because an excessive complexity leads to overfitting the training data.

This is a well known fact in linear regression, where a high-degree polynomial normally fits better to the data at the expense of the generalization capacity. As an illustration, Fig. 2.4 shows the fittings made with polynomials of degrees 3 and 15 to the function $\underline{y} = 1.5 + 1.3\sin(2\pi x) + \underline{\epsilon}$, where $\underline{\epsilon}$ is a Gaussian noise with zero mean and standard deviation equal to 0.1. It can be observed that while the polynomial of degree 3 is more or less equidistant from the samples, that of the order 15 tries to trace them all. This implies that in using the two polynomials different kinds of errors would occur. In the former case the error is known as *bias error*, since the fitted curve departs from the mean trend of the cloud of points. In the latter case, on the opposite, the trend is captured but there is a *variance error*, because the overfitting implies that the new samples will fluctuate randomly about the actual function that remains unknown. The ideal model corresponds to a trade-off solution between the two types of errors.

The above example illustrates the concept of *model complexity*. For one dimensional problems, a simple approach to its precise definition can be made in terms of the number of continuous derivatives of the function. For instance, in the example the polynomial of degree 15 is evidently much more complex than that of degree 3. This means that as the complexity grows there is a trend to overfitting. Consequently, the risk minimization should subject to control the model complexity and to select the simplest model possible. This idea has been given the name of Ockham's razor[3] and it is the basis of the regularization inductive principle described next.

The idea of *regularization* (also known as *penalization*) arose in the field of ill-posed problems [167] and is based on the Lagrange optimization. Recall that given the following constrained optimization problem

$$\begin{aligned} \text{minimize} \quad & J(\boldsymbol{x}) \\ \text{subject to} \quad & f_1(\boldsymbol{x}) < 0 \\ & f_2(\boldsymbol{x}) < 0 \\ & \vdots \end{aligned} \tag{2.18}$$

the solution can be found by creating the Lagrange functional (see e.g. [89])

$$W(\boldsymbol{x}) = J(\boldsymbol{x}) + \sum_i \lambda_i f_i(\boldsymbol{x}) \ , \tag{2.19}$$

[3] To honor the medieval philosopher William of Ockham (1298? - 1350?), whose critique to the proliferation of entities in the scholastic philosophy is summarized in the sentence *entia non sunt multiplicanda praeter necessitatem.*

where λ_i are called as *Lagrange multipliers*. This method was adopted by [167] after the observation that the penalization of a complex model in minimizing the learning risk can be interpreted as a bound in a general constrained optimization task. Hence, the regularization approach consists in choosing the function $h(\boldsymbol{x}, \boldsymbol{\omega})$ that minimizes the compound risk

$$R_{\mathrm{r}}(\boldsymbol{\omega}, n) = R_{\mathrm{e}}(\boldsymbol{\omega}, n) + \lambda \Psi \left(h(\boldsymbol{x}, \boldsymbol{\omega}) \right) , \tag{2.20}$$

where $\Psi \left(h(\boldsymbol{x}, \boldsymbol{\omega}) \right)$ is a non-negative functional, known as *regularizer*, in which the a priori knowledge on how models $h(\boldsymbol{x}, \boldsymbol{\omega})$ become complex is codified and $\lambda > 0$ is a penalization parameter that controls the relative weight of the risk. It is computed over the given samples and the approximating function. For instance, if λ is low the minimization risk would depend basically on the samples while a high value transfers the responsibility to the functional. As an illustration, a regularization approach for the simple one-dimensional polynomial regression is

$$\begin{aligned} R_{\mathrm{r}}(\boldsymbol{\omega}) &= R_{\mathrm{e}}(\boldsymbol{\omega}) + \lambda \Psi \left(h(x, \boldsymbol{\omega}) \right) \\ &= \frac{1}{n} \sum_{i=1}^{n} (y_i - h(x_i, \boldsymbol{\omega}))^2 + \lambda \int \left[\frac{\mathrm{d}^2 h(x, \boldsymbol{\omega})}{\mathrm{d}x^2} \right]^2 \mathrm{d}x \, . \end{aligned} \tag{2.21}$$

because the second derivative accounts for the curvatures of the model (see Fig. 2.4). The optimal value is normally calculated with repeated sampling techniques, such as the cross validation[4]

[175] argues that, unlike ERM, regularization is a non self-evident inductive principle, whose emergence was one of the first signs of intelligent inference. In particular, its application to the solution of ill-posed problems by [167] demonstrated that whereas self-evident methods of risk minimization do not always work, this non self-evident method does. This technique has been successfully used for improving the learning capacity of neural networks in several forms. Since common functions in structural mechanics are generally smooth (an a priori knowledge), the penalization of model complexity is a must in this field.

Regularization provides a general framework that underlies several procedures applied in statistical learning such as those known as *weight decay, early stopping, curvature smoothing, Bayesian inference, training with noise*, etc [16]. Let us examine some of them:

- *Bayesian Inference*. This is a technique applied in regression analysis that is based on the Bayes formula on the conditional probability for two random events A y B (see e.g. [15]):

[4]Cross validation is a procedure for selecting the best model from a family without increasing the sample set. It consists in training a member of the family successively with portions of the sample set removed and averaging the generalization error computed over the removed part. The selected model is that giving the lowest average error (see e.g. [69]).

$$P[B|A] = \frac{P[A|B]P[B]}{P[A]} . \tag{2.22}$$

This equation is normally used for all kinds of problems in which the inference starts from an initial (a priori) hypothesis on the probabilistic structure of the model ($P[B]$) before updating it ($P[B|A]$) after receiving new information ($P[A]$). For the learning problem that interests us, the Bayes formula takes the form

$$P[\text{model}|\text{samples}] = P[\text{samples}|\text{model}]\frac{P[\text{model}]}{P[\text{samples}]} . \tag{2.23}$$

Let us consider in particular the regression problem examined previously:

$$y = h(\boldsymbol{x}, \boldsymbol{\omega}^*) + \epsilon . \tag{2.24}$$

The training set is $\boldsymbol{q} = [\boldsymbol{X}, \boldsymbol{Y}]$, where $\boldsymbol{X} = [\boldsymbol{x}_1, \boldsymbol{x}_2, \ldots, \boldsymbol{x}_n]$ and $\boldsymbol{Y} = [y_1, y_2, \ldots, y_n]$. Let us suppose that the set of functions $h(\boldsymbol{x}, \boldsymbol{\omega}), \boldsymbol{\omega} \in \Omega$ contains the true function $h(\boldsymbol{x}, \boldsymbol{\omega}^*)$. Bayes formula gives then the posterior density of the probability that the set of parameters $\boldsymbol{\omega}$ corresponds to the target function[5]:

$$p_{\underline{\boldsymbol{\omega}}|\underline{\boldsymbol{q}}}(\boldsymbol{\omega}|\boldsymbol{q}) = P[\boldsymbol{q}|\boldsymbol{\omega}]\frac{p_{\underline{\boldsymbol{\omega}}}(\boldsymbol{\omega})}{P[\boldsymbol{q}]} , \tag{2.26}$$

where the probability that data $\boldsymbol{q}$ obeys model $\boldsymbol{\omega}$ can be taken as

$$P[\boldsymbol{q}|\boldsymbol{\omega}] = P[\boldsymbol{X}]\prod_{i=1}^{n} p_{\underline{\epsilon}}(y_i - h(\boldsymbol{x}_i, \boldsymbol{\omega})) , \tag{2.27}$$

due to the fact that the error ϵ can be assumed to be independent from data $\boldsymbol{X}$. Upon substituting (2.27) into (2.26) and taking logarithms it is obtained

$$R_{\text{b}}(\boldsymbol{\omega}, n) = \sum_{i=1}^{n} \ln p_{\underline{\epsilon}}(y_i - h(\boldsymbol{x}_i, \boldsymbol{\omega})) + \ln p_{\underline{\boldsymbol{\omega}}}(\boldsymbol{\omega}) . \tag{2.28}$$

If it is assumed that the error is Gaussian distributed, with variance $\sigma_{\underline{\epsilon}}^2$ and null mean, then

$$R_{\text{b}}(\boldsymbol{\omega}, n) = -\frac{1}{n}\sum_{i=1}^{n}(y_i - h(\boldsymbol{x}_i, \boldsymbol{\omega}))^2 + \frac{2\sigma_{\underline{\epsilon}}^2}{n} \ln p_{\underline{\boldsymbol{\omega}}}(\boldsymbol{\omega}) . \tag{2.29}$$

[5]Use is made here of the hybrid form of the Bayes formula:

$$p_{\underline{x}|\underline{y}}(x|y) = \frac{p_{\underline{y}|\underline{x}}(y|x)P_{\underline{x}}(x)}{P_{\underline{y}}(y)} . \tag{2.25}$$

Evidently, the Bayesian inference is equivalent to a regularized minimization of the risk under the Gaussian assumption. The main difference between both methods lies in the role of the coefficient λ, which now has an explicit value given in terms of the error variance, which may be estimated from the data. This implies that the Bayesian approach does not require the cross validation procedure mentioned above, which raises the computational cost. Notice that the a priori knowledge is also codified in the penalized term as before.

- *Weight Decay.* This technique is applied in training neural networks. It consists in minimizing the compound risk

$$R_{\mathrm{r}}(\boldsymbol{\omega}) = R_{\mathrm{e}}(\boldsymbol{\omega}) + \frac{\lambda}{2}\boldsymbol{\omega}^{\mathrm{T}}\boldsymbol{\omega} , \qquad (2.30)$$

which penalizes large values of the network parameters $\boldsymbol{\omega}$. The name of the method obeys to the fact that a sequential method of training multi-layer perceptrons could be devised in the form (see 4.27)

$$\boldsymbol{\omega}[k+1] = \boldsymbol{\omega}[k] + \eta\boldsymbol{\omega}[k] , \qquad (2.31)$$

whose continuous analogue is the differential equation

$$\frac{\mathrm{d}\boldsymbol{\omega}}{\mathrm{d}\tau} = \lambda\boldsymbol{\omega} \qquad (2.32)$$

having the solution

$$\boldsymbol{\omega}(\tau) = \boldsymbol{\omega}(0)\exp(-\lambda\tau) . \qquad (2.33)$$

Depending on the value of the penalization parameter λ, there is an effective number of parameters that is less than the number of interconnection weights, meaning that the complexity of the network is implicitly reduced [16].

- *Early Stopping.* This is a particular technique for training the class of neural networks known as Multi-Layer Perceptrons. Since these networks have the property of being universal approximators, their training error has a trend to diminish with a growing number of iterations K. However, their generalization error starts to grow after a certain number of training cycles (Fig. 2.5), which can be considered as the start of the overfitting. This justifies the early stopping of the training at an optimal iteration value K^*, just when the generalization error begins to grow. Evidently, the existence of a validation set implies an additional number of solver calls in a structural mechanics context, thus increasing the overall computational cost. To avoid this use can be made of the method of cross-validation mentioned in the preceding From the point of view of the inductive principles, it can be demonstrated that the early stopping procedure is equivalent to the weight decay method and hence to a regularization approach [16].

The regularization inductive principle attempts to overcome the simple ERM that dominates classical Statistics. For the field of structural reliability the application of the ERM implies either to search a model by trial and error that leads to an error reduction – thus augmenting the computational cost of the training phase – or to increase the number of samples, which implies additional calls of a finite element solver. Consequently, the application of the regularization principle is attractive from the point of view of computational stochastic mechanics. However, it is necessary to examine closely a more ambitious principle which is automatically applied by Support Vector Machines. This principle is known as *Structural Risk Minimization* (SRM) [174].

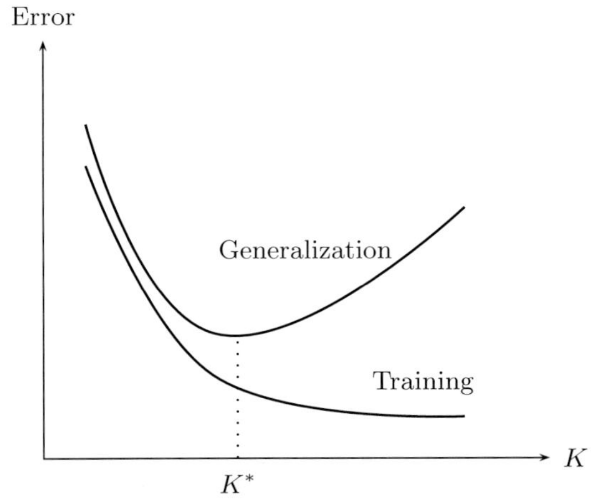

Fig. 2.5. Early stopping in training Multi-Layer Perceptrons (Adapted from [16])

In the regularization approach an attempt is made to reduce the empirical risk while keeping fixed the penalization term, because the model is assigned a priori. In the SRM the goal is to reduce both terms simultaneously by incorporating a model optimization procedure in the training process. However, before describing this principle it is necessary to introduce first some important concepts.

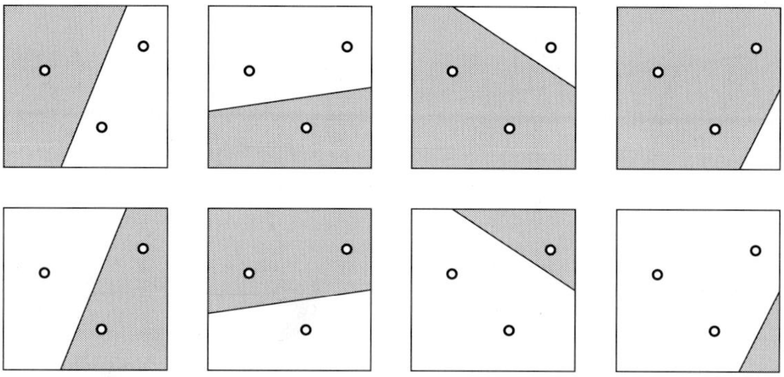

Fig. 2.6. Feasible dichotomies in a two-dimensional plane that can be realized by a linear function over three samples. There are 2^3 possibilities, thus indicating that for two dimensions the Vapnik-Chervonenkis dimension of a linear model is three (Adapted from [148])

2.5 Complexity and Vapnik-Chervonenkis Dimension

In the previous section the concept of complexity was introduced with reference to the simple case of fitting a polynomial function in the unidimensional case. For several dimensions, however, the definition of the complexity in terms of the number of derivatives leads to difficulties as it is obvious. In this section a more rigorous definition of complexity is summarized. The exposition is limited to the classification problem as this constitutes the basic learning task over which complexity measures for regression and density estimation can be construed.

Let us have two samples in the plane, which can belong either to a class labeled with a + or to the class ×. This means that the number of dichotomies is $2^2 = 4 : (+,+), (+,\times), (\times,+), (\times,\times)$. Advancing one step further in the induction, let us consider three samples, as shown in Fig. 2.6. If we were to separate these samples into the two classes by means of straight lines there would be a total number of 2^3 dichotomies, which are those illustrated in the figure. However, when the number of samples is four, the number of dichotomies is $2^4 = 16$ and the separation can no longer be performed by straight lines, as shown by Fig. 2.7. Thus a nonlinear separation is required.

On the basis of this example the *growth function* and the *Vapnik - Chervonenkis dimension* can be explained [176, 177]. Let us have a sample set $\boldsymbol{X}_n = (\boldsymbol{x}_1, \boldsymbol{x}_2,$

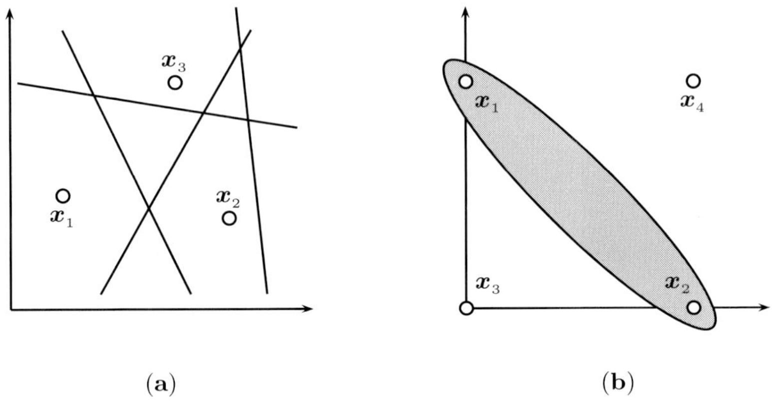

Fig. 2.7. On the Vapnik-Chervonenkis dimension. The lines in two dimensions can realize any combination of two classes if the number of samples is three. However, they cannot solve a logical XOR (Exclusive-Or) problem requiring four samples and a nonlinear decision function

$\ldots, \boldsymbol{x}_n)$. The total number of possible dichotomies that can be realized by the model is $N(\boldsymbol{X}_n)$. The model in the classification case is simply a member of the set of indicator functions $\Upsilon(Q(X_n, \boldsymbol{\omega}) > 0)$ of functions $Q(\boldsymbol{X}_n, \boldsymbol{\omega})$, each of which divides the space into two sectors. The growth function is defined as

$$G(n) = \ln \max_{\boldsymbol{X}_n} N(\boldsymbol{X}_n) , \tag{2.34}$$

where the maximum is taken over all the possible populations of size n. From the introducing example it is clear that for the separation with straight lines

$$G(n) \leq n \ln 2 , \tag{2.35}$$

where the inequality holds strictly for the separation with $n > 3$, because it is not possible to reach a perfect separation with the lines for more than three samples. On the contrary if a non linear model could discriminate n samples in the plane in all possible ways for any n, the growth function would increase monotonously.[6]

[6][175] derives philosophical consequences from this fact upon the basis of Popper's discrimination of scientific and non-scientific theories [128]. According to Popper, a theory cannot be said to be scientific if it cannot be falsified, i.e. if it can explain all phenomena. In like manner a functional model that could *explain* all

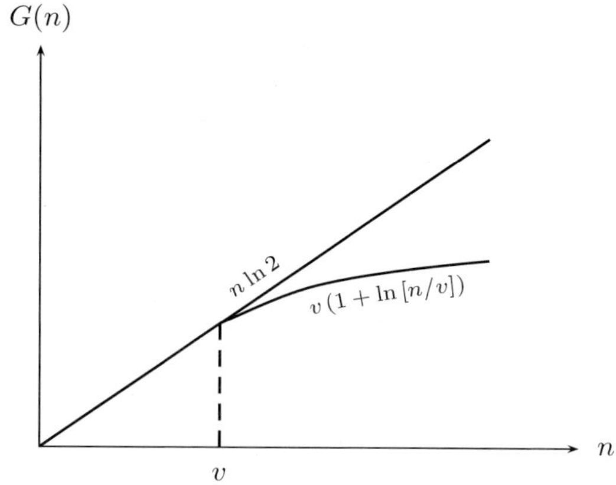

Fig. 2.8. Growth function (Adapted from [174])

On the other hand, the Vapnik-Chervonenkis dimension (briefly, VC-dimension) of the set of indicator functions $\Upsilon(Q(\boldsymbol{X}_n, \boldsymbol{\omega}) > 0)$ is defined as the maximum number v of samples that can be separated in all the 2^v ways that the model allows. Evidently $v = 3$ for the straight lines in the plane, according to the comments to Fig. 2.7. Generalizing, $v = d + 1$ for linear models, where d is the number of dimensions. The relationship between the growth function and the VC dimension is given by a theorem which states that the growth function is either linear ($G(n) = n \ln 2$) or bounded by $v\,(1 + \ln[n/v])$ (Fig. 2.8). The point defining the separation between the two functions is the VC dimension. This parameter has been calculated for several models employed in regression and classification [27, 4].

An important characteristic of the VC dimension is that it is not dependent on the probability density function of the samples. This allows the derivation of distribution-free error bounds in minimizing the classification risk. This is in contrast to the classical methods of statistical classification, which are based on Bayesian concepts and hence require postulating such

classes (i.e., that could realize all possible dichotomies and hence giving a null empirical risk) is the analogous of a non-scientific theory. The analogy is justified by the fact that both scientific theories and statistical models are construed by induction upon empirical facts over which an ideal (i.e. mathematical) model is imposed.

densities. These bounds are exposed in the next section. It is important to note before that the learning characteristics of regression functions such as neural networks can be more clearly defined in terms of other complexity measures known as *pseudo-dimension* and *fat-shattering dimension* [4]. Even a summary description of these involved parameters is beyond the scope of the present book, as it requires several additional concepts. However, some essential features of the complexity of neural networks can be described in terms of the VC dimension, as shown in Chap. 3.

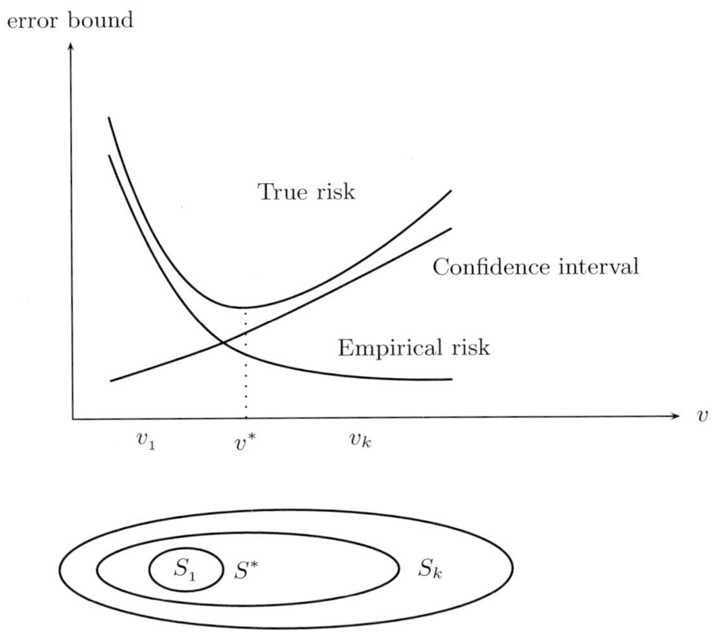

Fig. 2.9. Structured risk minimization and Vapnik-Chervonenkis dimension (Adapted from [27])

2.6 Error Bounds and Structured Risk Minimization

After introducing the above concepts we are in a position to complete the summary of the inductive principles with a brief description of the SRM. In a series of theorems, [174] demonstrated the existence of several bounds for the generalization risk of classification and regression models defined through their VC-dimension. In particular, he established that the following bound holds with probability $1-\eta$ for the binary classification risk with a set of functions $Q(\boldsymbol{X}_n, \boldsymbol{\omega})$:

$$R(\boldsymbol{\omega}) \leq R_{\mathrm{e}}(\boldsymbol{\omega}, n) + \frac{q}{2}\left(1 + \left[1 + \frac{4R_{\mathrm{e}}(\boldsymbol{\omega}, n)}{q}\right]^{\frac{1}{2}}\right) , \tag{2.36}$$

where

$$q = \frac{\alpha v[\ln(\beta n/v) + 1] - \ln(\eta/4)}{n} \tag{2.37}$$

and the parameters α and β are in the ranges $0 < \alpha \leq 4$ and $0 < \beta \leq 2$. Let us analyze (2.36). It indicates that if the model has an infinite dimension v (and hence it is non-falsifiable, according to Popper's criterion), then the risk is unbounded. This, for instance, occurs to the indicator function $\mathrm{sgn}[\sin(\omega x)]$, in which the frequency parameter ω can be chosen as little as desired to accommodate for any binary pattern on a line. On the contrary, if the VC dimension is finite then the risk has a finite bound that, as shown by the equation, does not depend on the underlying probability density function of the samples because v as well as $R_{\mathrm{e}}(\boldsymbol{\omega}, n)$ are distribution-free. This implies that the classification methods based on the control of the VC dimension (i.e. controlling the complexity) do not require hypothesis on the probability structure of the variables $(\underline{\boldsymbol{x}}, \underline{y})$, at a difference to classical Bayesian and parametric methods. This allows avoiding to solve a more complicated problem (i.e. the estimation of the class-probability densities) as an intermediate step in solving the target one (the classification), as recommended by [175][7].

An additional consequence of (2.36) is the possibility of minimizing the risk with a few samples by making the VC dimension the control variable. In fact, for a given VC dimension, if the ratio n/v is large (say, $n/v > 20$, which implies a large number of samples) then q is little and so is the second summand in (2.36). In such a case it is sufficient to minimize the empirical risk, as is commonly done in Statistics when there is no limitation on the number of samples. However, under such limitations (which are decisive in using these methods in structural reliability, where each sample implies e.g. a finite element solution) the value of q will not be necessarily little, thus imposing the need of reducing both terms at a time in order to minimize the risk. Consequently, instead of augmenting the number of training samples n it is necessary to make the VC dimension the variable that controls the empirical risk reduction in both summands of the equation as well as the value of q. This can be reached by searching the VC dimension that makes the sum a minimum.

Now, when increasing the VC dimension the empirical risk diminishes because the model complexity grows and hence its fitting to the training data. However, by the same token the generalization error also grows and hence

[7] In a sense, this recommendation follows from Ockham's economy principle recalled in the preceding.

the confidence interval of the model because the variance error increases. In conclusion it is necessary to reach a trade-off solution, as is usual in Statistics when dealing with the bias-variance dilemma. The novelty in this case is the condition of using a low number of samples. This can be solved with the inductive principle known as *Structural Risk Minimization* that is explained next.

Let S be a set of functions $Q(\boldsymbol{X}_n, \omega), \omega \in \Omega$, equipped with a nested set structure $S_k = \{Q(\boldsymbol{X}_n, \omega), \omega \in \Omega_k\}$, i.e.

$$S_1 \in S_2 \in \ldots \in S^* \in \ldots , \tag{2.38}$$

such that

$$v_1 \leq v_2 \leq \ldots \leq v^* \leq \ldots . \tag{2.39}$$

The goal is to seek the value of k yielding the minimal bound of the expected risk, consisting in the sum of the empirical risk and the confidence interval (see (2.36) and Fig. 2.9).

The consequence of reaching such a minimum is evident: One would obtain a model trained with a few samples but with the maximal generalization capacity, within the generalization capabilities allowed by such number of samples. In comparison to other methods and principles mentioned above there will be the following advantages:

- There is no need of a validation set, which in stochastic mechanics requires additional solver calls.
- The cross-validation, that largely increases the training computation time, is also not required.
- At a difference to Bayesian and parametric methods, the SRM is distribution-free and hence it does not require the estimation of densities.
- Training can be performed with a few samples because the VC-dimension controls the reduction of the empirical risk.

2.7 Risk Bounds for Regression

As said in the preceding, the implicit reliability problem can be faced using solver-substitution techniques in two forms, namely regression or classification. The latter has the advantage that the overall risk is less sensible to the selection of a suboptimal model. In fact, in the regression case, the bound of the overall risk takes the form [174]

$$R(\boldsymbol{\omega}) \leq \frac{R_{\mathrm{e}}(\boldsymbol{\omega}, n)}{(1-\sqrt{q})_+} , \tag{2.40}$$

where $(p)_+ \equiv \max(0, p)$ and $R_{\mathrm{e}}(\boldsymbol{\omega}, n)$ is the empirical risk, which in RSM, neural network training procedures and other regression techniques is the conventional least squares:

$$R_e(\boldsymbol{\omega}, n) = \frac{1}{n}\sum_{i=1}^{n}[g(\boldsymbol{x}_i) - \hat{g}(\boldsymbol{x}_i, \boldsymbol{\omega})]^2 , \qquad (2.41)$$

where $g(\boldsymbol{x}_i)$ and $\hat{g}(\boldsymbol{x}_i, \boldsymbol{\omega})$ are the actual and postulated performance functions, respectively. It is evident that a suboptimal model in (2.40) has a multiplicative effect on the risk, while merely an additive effect in the bound (2.36) for the classification risk. This comparison reflects the intuitive appreciation that the estimation of the value of a function is more sensible to errors in model selection than that of its sign.

This difference is particularly important to some classes of neural networks, since they can be used either as regressors or as classifiers. Specifically, Multi-Layer Perceptrons using nonlinear activation functions in the output layer require nonlinear optimization techniques, which can easily become trapped in a local minimum, corresponding to a suboptimal model. In such situation, their use as classifiers is, at least theoretically, more reliable than as regressors according to this comparison. On the contrary, Support Vector Regression operates a linear mapping to the so-called feature space, so that there is no risk of a suboptimal solution.

2.8 Stringent and Adaptive Models

Another, very important issue that is worth consideration is that of the structure of the learning model in terms of adaptivity and flexibility. These notions were introduced in the previous chapter but are now considered in a wider perspective. It was shown there that classical learning methods such as polynomial regression are characterized by constituent functions that are not indexed by the samples. In other words, they use functions such as $x, x^2, \cos(x)$, etc., which do not depend on the sample position and hence have infinite support. Modern learning methods, on the contrary, employ functions that in some way depend on the sample coordinates and hence the functions or their derivatives have finite active support. This situation is clearly exemplified in function approximation, where the harmonic basis used in the Fourier representation have infinite support, which is in contrast to the constituent functions employed in wavelet analysis[8].

Adaptive models have been defined as those composed by a weighted sum of functions whose parameters are determined by the given samples. For instance, the general formula

$$\sum_i w_i h(\boldsymbol{x}_i, \boldsymbol{x}, \boldsymbol{\omega})) \qquad (2.42)$$

holds for kernel and finite mixtures for density estimation, some kinds of Neural Networks used for regression as well as for classification, Support Vector

[8]The wavelet transform is summarized in the next chapter.

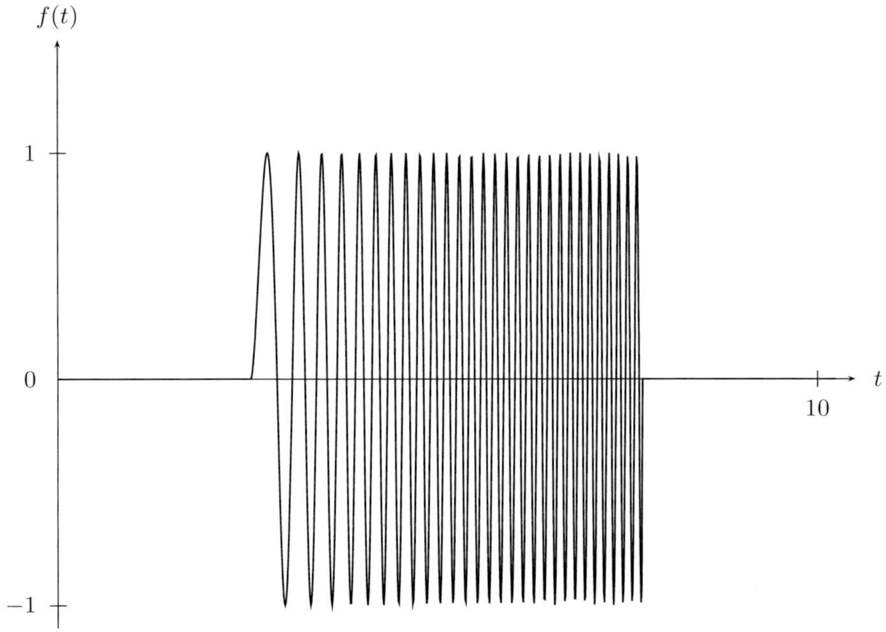

Fig. 2.10. Example signal for wavelet analysis

Machines, wavelets (see e.g. [108]), fuzzy regression approaches [87], etc. In this equation $\boldsymbol{x}_i$ are the sample vectors, w_i the weighting coefficients, $\boldsymbol{\omega}$ a set of parameters and $h(\cdot)$ a nonlinear function. A high *flexibility* of the model can be obtained if function $h(\cdot)$ or its derivative has a localized effective support (i.e. the support for which $h(\boldsymbol{x}) < \epsilon$ or $h'(\boldsymbol{x}) < \epsilon$, for an arbitrary small parameter ϵ). An example of the former is the Radial Basis Functions used by some of the models mentioned above, while the classical example for the latter is the logistic sigmoid function used in Multi-Layer Perceptron. On the other hand, the *adaptivity* is a consequence of the dependence of the function on the samples $\boldsymbol{x}_i$.

It is well established that these two characteristics of approximation methods render positive effects in reducing the error of the representation of the underlying (i.e. true) function with the model. This is because the finite support of the function implies that an error occurring in some position does not have an effect over the entire domain of the approximation. On the contrary, in models using functions with infinite support, the error in the representation is, so as to say, transmitted through the entire domain thus affecting the

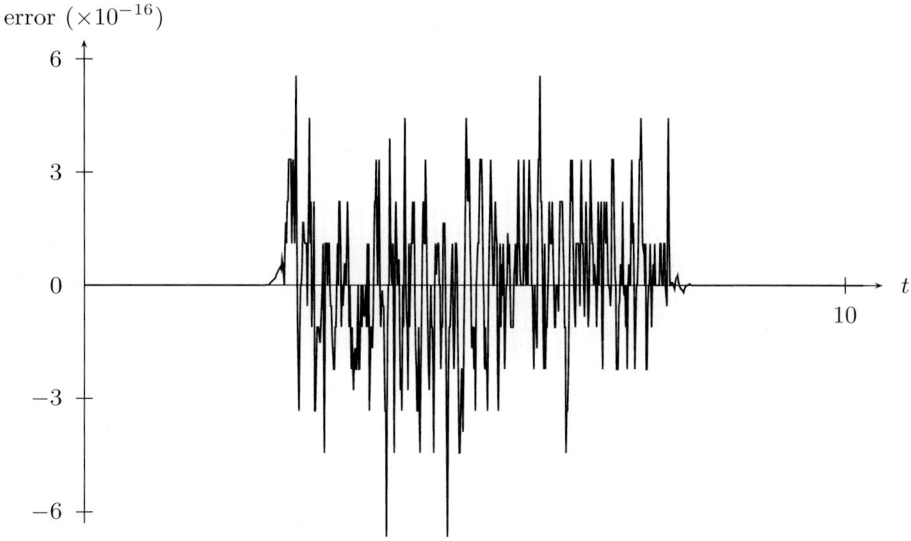

Fig. 2.11. Error in the reconstructed signal using wavelet transform

entire representation. This is illustrated by Figs. 2.10 to 2.12. The first shows the example signal, which raises suddenly from zero and varies with a modulated frequency before disappearing also abruptly. The other two figures show respectively the error corresponding to the inverse transforms in the Fourier and wavelet cases[9]. The benefits of the adaptivity of the wavelet method are evident not only because the errors are of lower order but more importantly because the error signal does not extend beyond the domain of the function, in contrast to the Fourier representation.

2.9 The Curse of Dimensionality

One of the most important problems that must be faced in structural reliability as well as in statistical learning is that concerning the effects of the growing of dimensions in which the problem at hand is specified. In both cases the number of dimensions is given by the size of the random variable

[9]The Fourier analysis was performed with 1024 points. A Daubechies wavelet of order 4 was used in the calculation.

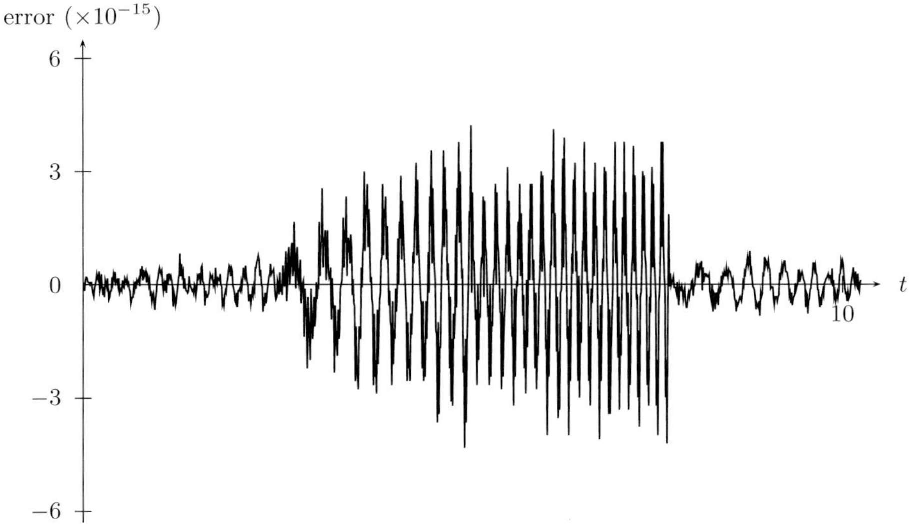

Fig. 2.12. Error in the reconstructed signal using Fourier transform

set, which is denoted as d in this book. These problems are known collectively as the *curse of dimensionality*, an expression coined by [13].

The applicability of some classical methods of stochastic mechanics is limited by this parameter. In fact, in using second order methods such as FORM or SORM for implicit limit state functions it is necessary to compute their gradient and/or the derivatives of the stiffness matrix with respect to the random variables composing the input set [97, 63]. This kind of computations implies an explosion of dimensions since these quantities are by definition Kronecker products. A similar increase is obtained in applying the spectral method based on Karhunen-Loève expansion and polynomial chaoses [55]. The situation is worse for stochastic dynamics problems in which the number of variables d is usually very large.

If the probabilistic information on the basic variables is sufficient for a Monte Carlo simulation, this method can be preferable in cases involving a large number of basic variables, because Monte Carlo integration as such is scarcely sensitive to the dimensionality, as said in Chap. 1. However, due to the randomness of the failure probabilities in Monte Carlo methods, the number of solver calls should be augmented in order to reduce the confidence interval

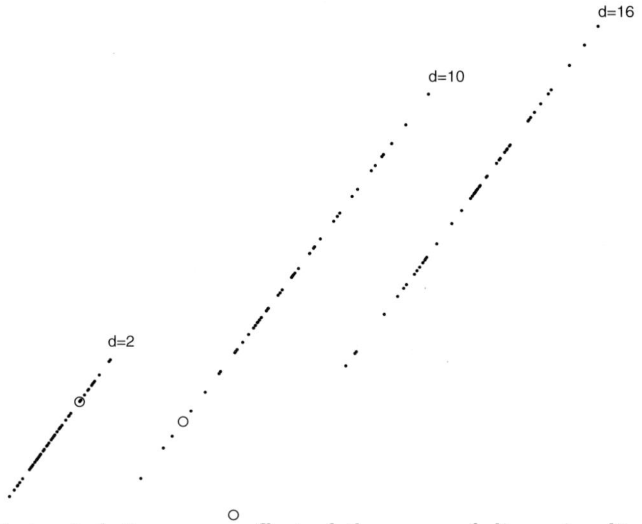

Fig. 2.13. Vector isolation as an effect of the curse of dimensionality. The dots represent the projection of uniform samples in $d-$dimensional spaces on the vector $(0.5, 0.5, \ldots)$, appearing as a circle

of the estimate. This makes valuable the application of solver-substitution techniques, which is the main motto of this book. Since these techniques consist in the computation of a regressor or a classifier after some available samples, it is important to examine the dimensionality effects on learning from them.

The main effects of dimensionality growing on random populations of samples are the following [27]:

- As the dimensionality grows the number of samples required to yield the same density increases exponentially. Put otherwise, a larger radius is needed to enclose the same fraction of samples. To be explicit, in a hypercube of unit length and d dimensions, the expected edge length of a smaller hypercube needed to capture a fraction r of the samples is given by $r^{1/d}$. As an example, for $d = 2, 10, 16$ it is necessary to cut the unit hypercube at 0.1, 0.63 and 0.75 of its edges in order to capture only one percent of the data. This explains the problem posed by curvatures of the same sign in reliability analysis which was discussed in Chap. 1.

- The mean distance among points increases and almost every point is an outlier in the projection of all points upon its unit vector. This is illustrated by figure 2.13 for same dimensionality degrees as above. In the particular case of multivariate standard Normal density, which is of high importance in structural reliability, it has been demonstrated that the expected value Euclidean distance from the origin to a sample grows linearly with the number of dimensions [9].
- Finally, the samples tend to be close to an edge of the population. In particular, for n points uniformly distributed in a unit hypershpere, the median distance from the origin to the closest point is

$$\rho(d,n) = \left(1 - \frac{1}{2}^{\frac{1}{n}}\right)^{\frac{1}{d}} . \tag{2.43}$$

For $d = 2, 10, 16$ and $n = 1000$ this yields $\rho = 0.026, 0.483, 0.636$, meaning that, as the dimensionality increases, the samples are the closer to the border of the hypersphere, which in turn has a higher hypersurface than any other ball contained in the sphere.

The way in which this problem affects the main learning methods that are useful in structural reliability will be discussed in Chaps. 3 to 6. However, there is an specific situation (namely, the classification problem) in which the increase of the number of dimensions renders positive effects. This fact, which is explained next, is exploited by Support Vector Machines and other kernel methods, as discussed in Chaps. 3 and 5.

2.10 Dimensionality Increase

The dimensionality increase, which creates difficulties in solving many mathematical and numerical problems, renders positive effects in the specific statistical field of pattern recognition. In fact, the curse of dimensionality is somewhat reversed in this case, as the dimensionality increase facilitates discriminating the classes by means of a simple hyperplane. This fact is incorporated in the support vector method of classification.

To be specific, a characterizing feature of hyperplane discrimination is that the probability of correct classification of n samples increases with the number of dimensions d. In fact, the probability of correct hyperplane classification of a two-class training set of size n and input dimensionality d is [47]

$$P(n,d) = \begin{cases} 1, & n \leq d+1 \\ \left(\frac{1}{2}\right)^{n-1} \sum_{k=0}^{d} \binom{n-1}{k}, & n > d+1 . \end{cases} \tag{2.44}$$

This is based on a theorem by [31]. Figure 2.14 illustrates the distribution (2.44) for three training set sizes, which can be regarded as economic for

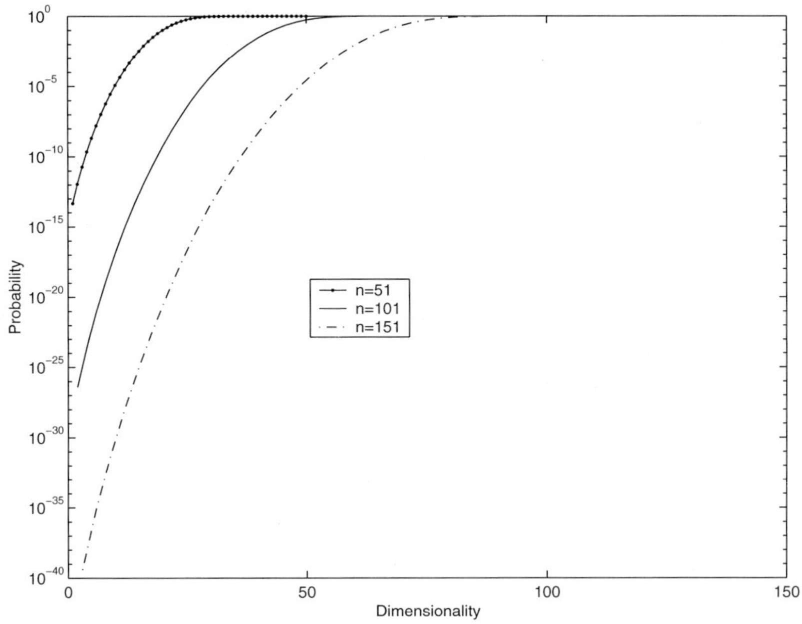

Fig. 2.14. Probability of correct classification with a hyperplane as a function of the dimensionality

many structural reliability analysis problems. It can be observed that beyond sixty dimensions an almost perfect binary classification is assured for $n = 101$. Put another way, if $n/d < 2$ there is a very high probability that the samples are correctly classified by a hyperplane, and if $d = n - 1$ the right classification is granted. This is more clearly illustrated by figure 2.15 which shows the relationship of the probability of right classification as a function of the parameter $n/(d+1)$.

A condition for this to hold is that the samples are in the so-called *general position*, defined as follows: We say that n points in $\mathbb{R}^d$ are in general position if for no k, such that $2 \leq k \leq d+1$, do k of these points lie in a hyperplane of dimension $k-2$ [47]. In words, since a $k-$dimensional hyperplane is determined by $k+1$ points, the definition requires that no more than $k+1$ points are in a hyperplane. This condition is easily met by nonlinear mappings used in the support vector methodology for increasing the dimensionality. Also observe that a complete lack of correlation among the basic variables in a structural reliability problem is the best situation for generating samples in general position, because in that case there is a minimal possibility of linear

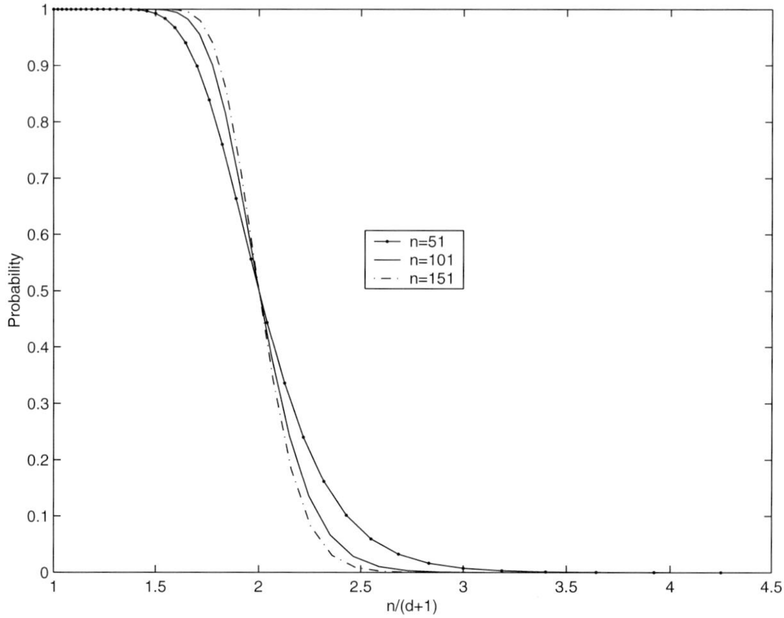

Fig. 2.15. Probability of correct classification with a hyperplane as a function of the ratio of sample size to dimensionality

dependence among a subset of samples. This favors the transformation of the reliability problem to the standard Normal space for applying these kind of learning machines using e.g. the Nataf method summarized in Chap. 1.

2.11 Sample Complexity

Further insight into the learning problem and, in particular, the dimensionality effects, can be obtained by considering the theory on the lower bound for the training samples, necessary to reach a prescribed error with a certain probability. This concept is known as *sample complexity* [4]. Rigorously, the sample complexity of a learning algorithm L is the smallest integer that is sufficient for having, with a probability higher than $1 - \eta$, a classification error e. This probability is written as

$$P[e < \inf(e) + \epsilon] \geq 1 - \eta \ , \qquad (2.45)$$

where $\inf(e)$ is the highest lower bound of the error that can be obtained with the given set of functions and $\epsilon > 0$ a parameter.

The lower bounds for the sample complexity $N_{\mathsf{L}}(\epsilon, \eta)$ that apply for binary classification in the structural reliability case correspond to the situation when there is a target, though unknown, function $g(\boldsymbol{x})$. In the nomenclature of Statistical Learning this situation is called a *restricted model*, which requires less training samples than the opposite case in which such a function does not exist [4]. For the restricted case, the sample complexity is bounded by below as

$$N_{\mathsf{L}}(\epsilon, \eta) > \frac{1}{2\epsilon} \ln \left(\frac{1}{\eta} \right) \tag{2.46}$$

for $0 < \epsilon < 3/4$ and $0 < \eta < 1$. Assuming for instance $\epsilon = \eta = 0.05$, the minimum sample size will be 30, regardless of the dimensionality of the design variable space. This is indeed an important result that fosters the search for parsimonious algorithms that exploit this learning capability of classifiers. However, it must be stressed that this by no means implies that such a low number of samples is sufficient when applying any learning device, because its structure can be so complex that the number of training samples will necessarily be very large. This complexity is controlled by the VC-dimension. In fact, for lower admissible errors and probabilities ($0 < \epsilon < 1/8$ and $0 < \eta < 1/100$) another practical result applies:

$$N_{\mathsf{L}}(\epsilon, \eta) \geq \frac{v - 1}{32\epsilon} , \tag{2.47}$$

where v is the Vapnik-Chervonenkis dimension.

The dimensionality effects on SL devices will be examined in subsequent chapters.

2.12 Selecting a Learning Method in Reliability Analysis

When facing the problem of creating a finite element (or similar) solver surrogate for solving structural reliability problems the analyst faces the problem of selecting a learning method among the many that have been published in the specialized statistical literature. In this section the applicability of SL techniques to the specific structural reliability problem is discussed. The examination is guided by the classification regard, as this problem underlies the regression and other approaches.

2.12.1 Classification Techniques

For the classification of samples we have a large list of alternatives [69, 148]:

- Linear Discriminant Analysis and Logistic Regression, which seek a separating hyperplane on the basis of Bayes formula.
- Rosenblatt's Perceptron, which pursues the same goal without resort to probability density information.
- Bayesian Classification, which requires the knowledge of the class-density functions or approximations thereof.
- Classification Trees, which builds a classifier as a collection of inequalities hierarchically ordered.
- Nearest Neighbor Algorithms, in which the classification rules are built on distance basis.
- Multi-Layer Perceptrons, which are Neural Networks equipped with threshold or sigmoid functions in their output units.
- Probabilistic Neural Networks, which are a flexible implementation of Bayesian classification.
- Support Vector Machines and their variants, which are kernel-based approaches that can be regarded as a development of Rosenblatt's perceptron for dealing with nonlinear separation.
- Bayesian kernel methods, which exploit the advantages of the dimensionality increase caused by the kernel formulation in a classical Bayesian framework.

In principle, any regression method can also be used for classification, since the latter can be treated but as a special regression case with binary or discrete outputs. This is the case, in particular, of the Logistic Regression mentioned in the list. For this reason, the list could be enlarged with adaptations of many regression techniques.

2.12.2 Remarks on Probability Density Estimation

An important area of statistical learning concerns the estimation of the probability density function in the univariate and multivariate cases. Starting from the work of Parzen in the sixties [124], some researchers have developed methods for this task aimed at improving the basic histogram estimate or at developing the method of kernel estimation proposed by Parzen for the univariate case. A great impulse in this direction was received from the invention of the *Expectation-Maximization* algorithm (EM) which allows to find the optimal parameters $\{w_j, v_j\}$ for the joint density model [34]

$$p_{\underline{\boldsymbol{x}}}(\boldsymbol{x}) = \sum_j w_j h(\boldsymbol{x}, v_j) \tag{2.48}$$

of the random variables $\underline{\boldsymbol{x}}$ on the basis of some available samples. In this equation $h(\boldsymbol{x}, v_j)$ is a kernel density function, which is most commonly the multivariate Gaussian. Much of the research on the theory and applications of modern density estimation methods is performed around this algorithm.

In spite these developments, it is important to point out that the density estimation problem as such is harassed by some difficulties:

- *It is an ill-posed problem* in the sense defined by [62]. In fact, the estimation of the density function in the univariate case requires the solution of the integral problem

$$\int_{-\infty}^{x} p_{\underline{x}}(x)\mathrm{d}x = P_{\underline{x}}(x) , \tag{2.49}$$

  where the distribution function is unknown and we have only a collection of some i.i.d. samples $x_1, x_2, \ldots x_n$. The ill-posed nature of the problem (2.49) stems from the fact that a small change in the right hand side of the equation implies large differences in the solution [175]. The situation, of curse, is worse in the multivariate case.
- *It is seriously affected by the curse of dimensionality.* This has already been illustrated in Chap. 1 by quoting an example discussed by [152].

The application of regularization techniques and other advanced inductive principles may reduce the effects of the ill-posed nature of the density estimation problem. The dimensionality issue, however, seems to be impossible to overcome, as it stems from the very spreading effects caused by the increase of the number of dimensions. In other words, the probability density function is just the mathematical expression of the sample spread in high dimensions. Taking into account that the number of dimensions in an actual stochastic mechanics problem is large, the learning methods for density estimation are not examined at a larger length in this book. In general, it can be said that reliability methods implying the estimation of a multivariate density function in moderate- to high-dimensional spaces should be avoided.

In spite of this, adaptive methods for density estimation can be useful as data compression techniques, which is a form of unsupervised learning that allows the selection of some relevant samples out of a large population. This is an interesting option for preprocessing the Monte Carlo samples that could be used in training statistical learning devices. For this reason Sect. 3.6.2 of Chap. 3 is devoted to the the Expectation-Maximization method for density estimation, which yields some parameters that can be used in that way.

2.12.3 Characteristics of Samples in Structural Reliability

All the classification methods mentioned in Sect. 2.12.1 have been developed in the field of Pattern Recognition (see e.g. [139, 166]), whose aim is to assign samples to labeled classes. In Structural Reliability, however, there are some specific features that influence the selection of the learning method. These are:

- *Known distribution.* At a difference to most pattern recognition tasks, the underlying distribution function of the samples in structural reliability is assumed to be known. This has important implications on the Bayesian regard to classification, as discussed in Chap. 4. Also, from the theoretical perspective, this characteristic implies that structural reliability problems belong to the category of the restricted models already defined. On the other hand, in contrast to the training approach adopted in the Response Surface Method, in which no strong use is made of the knowledge of the joint distribution, for statistical learning procedures this knowledge has an important practical consequence: the samples generated from the actual joint distribution naturally lead the learning device to the region in a high dimensional space where the training should be done. Such training can be performed by simply generating a large population of samples after the joint distribution, removing a large portion of them that lie in the safe domain and then reducing the resulting set with a data compression procedure, as detailed in see Chaps. 3 to 6.
- *Artificial samples.* In structural reliability practice the samples are rarely natural (i.e derived from experience) but are more often synthetic, so that their production can be managed for the sake of the better performance of the algorithms. The production of the samples is done on the basis of the distribution function.
- *Sample cost.* In structural reliability each sample is obtained from a numerical solver, so that in some situations its computational cost can be high and hence the learning method should be as economical as possible.
- *Perfect class separability.* In the classes observed in nature, it is often the case that they are either well separated or are merged in such a way that perfect separation is impossible. As an illustration, consider Fig. 2.16 which shows the two most important features of three varieties of the plant *Iris*. It is evident that while the *Iris setosa* forms a clearly distinguished class, there is no perfect separation between the *Iris virginica* and the *Iris versicolor.* As a consequence an error must be admitted in using a classification function in this case.

  On the other hand, the two classes in the specific problem of structural reliability are infinitely close to each other but, at the same time, are perfectly separable, due to the fact that they are determined by an underlying, deterministic performance function, which normally is monotonously decreasing, as illustrated by Figure 2.17. This characteristic makes the use of modern classification techniques easier than in most problems faced in pattern recognition, as the models need not be overly involved and no strong compromises should be made for balancing the bias and variance errors. Also, it has a decisive effect in selecting classification methods, because the perfect separability implies that statistical measures of the classes, on which classical methods are based, are of secondary importance with respect to their mutual boundary, which are the target of some modern

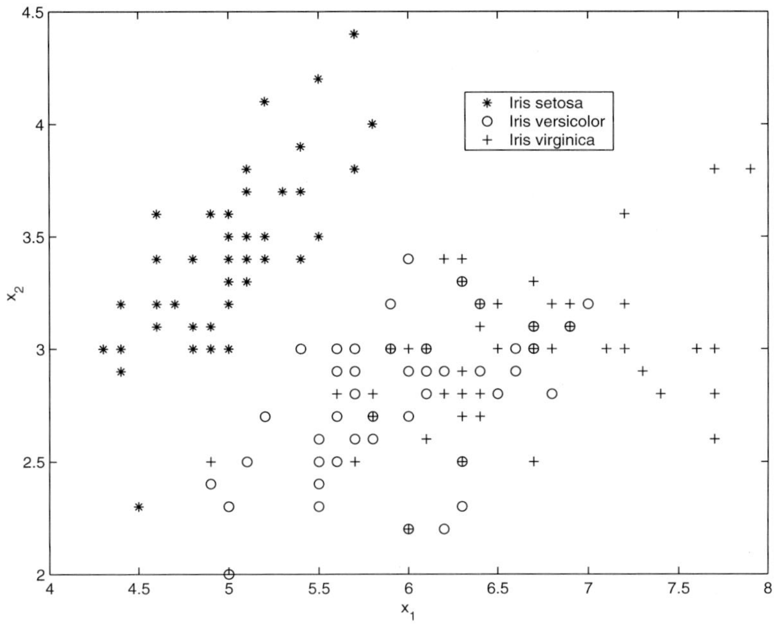

Fig. 2.16. Common discrimination patterns in samples drawn from nature. Features of the varieties of the *Iris* plant. (The data for drawing the plot have been taken from [109])

learning methods. In addition, some methods especially aimed at solving the non separable case are of little interest in this application.

- *Known intrinsic dimensionality.* In dealing with samples arising from nature it frequently happens that the actual dimensionality of the samples is less than that of the raw data and hence some preprocessing methods should be applied in order to reveal it. For instance, low correlation between the variables implies that the intrinsic and given dimensionalities are equal, whereas high correlation between two variables means that the intrinsic dimensionality is one degree less. However, for the samples generated by Monte Carlo simulation in structural reliability the actual dimensionality is known in advance, since the covariance matrix is given beforehand.

In selecting the learning method it should also be taken into account the nature of the stochastic problem with respect to the learning approach, for

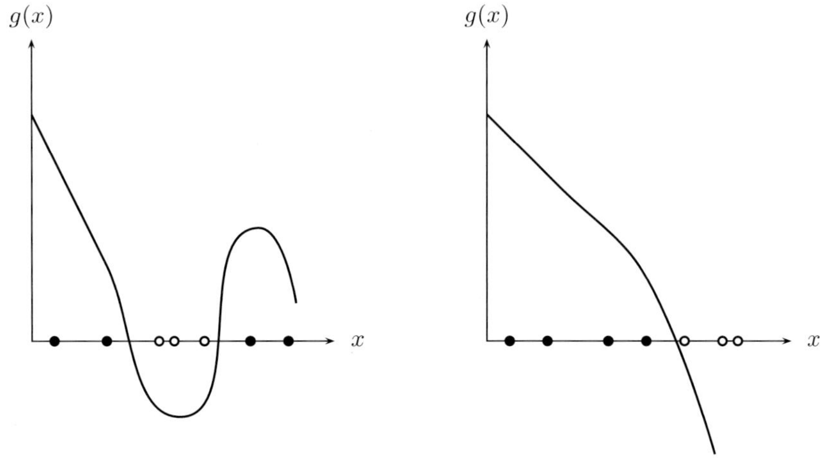

Fig. 2.17. Class separability in structural reliability analysis. (**a**): Imperfect separation (abnormal); (**b**): Perfect separation (normal)

the latter can require the solution of intermediate problems more involved than the target one. Let us consider two examples:

- The reliability problem can in principle be treated more simply as a classification problem rather than as a regression or a density estimation one, which seem rather more complicated. The discussion made on the the bounds of learning risks for classification and regression (2.36 and 2.40) illustrates this. However, this by no means suggests that any classification method is simpler than any other regression one, since, for instance, some classification techniques such as Bayesian Discrimination and Probabilistic Neural Networks require the estimation of class-density functions as intermediate steps, as will be shown later on. As said in the preceding, this is is an ill-posed problem highly affected by the curse of dimensionality. The differences in accuracy and computational effort between classification and regression using flexible methods such as neural networks are minimal in comparison to that existing between them and the methods requiring multivariate density estimation.
- A similar reasoning applies to the case of Monte Carlo simulation techniques, especially those requiring a density estimation in the way of calculating the failure probability, such as Optimal Importance Sampling

methods. This issue has been addressed in Chap. 1. Even with a good approximation of the performance or the limit-state functions, the calculation of the failure probability by means of the Importance Sampling estimate

$$\hat{P}_{\mathrm{f}} = \frac{1}{n} \sum_{i=1}^{n} \Upsilon_{\mathcal{F}}(\boldsymbol{x}_i) \frac{p_{\underline{\boldsymbol{x}}}(\boldsymbol{x}_i)}{h(\boldsymbol{x}_i)} \tag{2.50}$$

is a complicated problem from the point of view of accuracy, due to the severe effects of the curse of dimensionality on the estimation of an optimal $h(\boldsymbol{x})$ using a given set of samples.

2.12.4 A Look from the Viewpoint of Information Theory

It has been said in the preceding that, at a difference with respect to most problems faced in statistical learning, the training population in structural reliability is fundamentally synthetic. Such a population could then be generated in two forms, namely with deterministic or random plans. In contrast to the methods derived from Experimental Design, for training statistical learning devices it is preferable to make use of random populations for the following reasons.

For function approximation in general there underlies an optimization problem, namely that of minimizing the difference between the target and the hypothesized function. Also, from the reliability analysis viewpoint there is also an optimization task, which is that of finding the design point.

These references to the optimization problem are important for examining the effects of deterministic sample specification or random sample generation. In fact, the deterministic setting hampers the positive effects of random search models, such as Monte Carlo optimization [140], Genetic Algorithms [58] and Evolutionary Strategies [10], which have been successfully applied in solving complex optimization tasks. The rationale behind such positive effects is well explained by the thermodynamic analogy on which the evolutionary method known as Simulated Annealing is based [138] but in a deeper analysis they are explained by the Information Theory. In essence, the information increase (measured in bits) given by the occurrence of a random event A with a probability $P(A)$ is given by

$$I(A) = \log_2 \frac{1}{P(A)} \,. \tag{2.51}$$

In Information Theory a bit is defined as the amount of information received when one out of two equally probable alternatives is given, i.e. when $P(A) = 1/2$. Now, for a deterministic event A, $P(A) = 1$, $\log_2(1) = 0$ and hence $I(A) = 0$. In words, deterministic (or sure) events do not report an information increase. A set X of random samples $\boldsymbol{x}_i, i = 1, 2, \ldots, n$ used for building a solver surrogate offer the possibility of increasing the information by an increase of the entropy given by

$$H(X) = -\sum_{i=1}^{n} P(\boldsymbol{x}_i) \log_2 \left(\frac{1}{P(\boldsymbol{x}_i)} \right) , \tag{2.52}$$

whose thermodynamic analogue is a temperature increase that inputs energy into the system. For an optimization problem, this temperature raise implies an overcoming of local minima in which the optimization procedure may be trapped and the possibility of exploring new regions. This explains why the stochastic search methods mentioned above succeed in finding the solution of multiple problems of extreme values and, also, of inverse problems (see e.g. [164]) of which the building a functional approximation to the implicit, finite-element given functions used in structural reliability is but a particular instance. The practical meaning of these Information Theory ideas is that the sequential refining of a trial model can be more efficiently achieved by means of random (i.e. informative) samples rather than with deterministic, non-informative ones. An example of this efficiency in the field of structural reliability is given by adaptive Monte Carlo techniques based on Markov chains [7], given the connection between Markov theory of probability transition and random search concepts, as testified by the fact that both the Simulated Annealing optimization technique and the Markov Chain Monte Carlo (MCMC) method have a common origin in the seminal paper by [116]. In fact, the Simulated Annealing is but a MCMC in which the new states are accepted or rejected according to the Boltzmann probability distribution [140].

To end this digression on Information and Optimization viewpoints, a note on FORM and SORM is in order. As shown in the preceding, FORM is an attempt to solve the optimization problem of finding the design point and gives a surrogate for the limit-state function in the form of hyperplane as a byproduct. On the other hand, SORM is an improvement of such a surrogate that includes second order terms. Therefore, it can be said that FORM and SORM, despite being also rigid models as is evident after the discussion on the RSM, are in some sense superior to this latter in that they incorporate the optimization problem in building these approximating functions. However, the gradient and Hessian matrices are needed to these purposes. The stochastic methods for optimization mentioned above were developed just as attempts to overcome the need of calculating such information on derivatives, which in some cases is not at a disposal; for this reason, they are sometimes called gradient-free optimization methods [88]. In structural reliability, the gradients and curvatures are only available in some simple cases and are in general unknown when the calculation of the limit-state function depends on a finite-element solver. Hence, it is concluded that a combination of random-search optimization techniques with adaptive-flexible statistical learning devices would render two benefits:

1. To overcome the dependence of gradient and second-order derivatives for finding optimal solutions.
2. To exploit the adaptivity and flexibility of the models to overcome the rigidity of FORM, SORM and RSM models.

It is noteworthy that some recent proposals for calculating failure probabilities incorporate in some manner ideas of evolutionary strategy and learning from samples, without being explicitly drawn from such fields. These are, for instance, the method proposed by [132], in which a merit function is used for selecting the new samples, as in Genetic Algorithms, and the adaptation of the Metropolis method to structural reliability developed by [7], whose connection to Simulated Annealing has already been mentioned. This suggests that the incorporation of concepts and methods from the fields of Statistical Learning, Information Theory and Stochastic Optimization would be to large avail for Structural Reliability. Present monograph is oriented in such direction.

2.12.5 Recommended Methods

Statistical Learning devices such as MLP, RBFN and SVM can be applied either for regression or classification tasks. In the former case they can be used either as a method based on the performance function or on the limit-state function. The classification approach, however, can only be used for separating samples in the two domains and hence it is a method based on the limit-state function. The regression approach is convenient when the interest lies in estimating the entire distribution function of the observed response in (1.2), i.e. for methods based on the performance function. Table 2.1 indicates the estimates that can be obtained with the techniques that are recommended in parentheses. The recommendation obeys not only empirical observations but also some theoretical reasons that are detailed in the subsequent chapters.

Table 2.1. Applicability of Statistical Learning methods

Function	Classification (SVM, MLP)	Regression (SVM,MLP,RBFN)
Performance	——	$P_{\underline{g}}(g)$
Limit-state	P_{f}	P_{f}

3

Dimension Reduction and Data Compression

3.1 Introduction

The discussion made in the preceding chapter on the curse of dimensionality points out the need of the *dimensionality reduction* of the learning problem. Related to this are *feature extraction* techniques which transform the input data into another set of lower dimensionality that shows better the inner structure of the given population. In general, these techniques can be viewed as preprocessing methods. However, in many cases they can be useful also for postprocessing the output of a Monte Carlo simulation. There are also techniques for *data compression* i.e. the reduction of a population to a representative set of smaller size.

A well known unsupervised learning method for achieving the first goal is the *Principal Component Analysis* (PCA), which reduces the number of random variables of interest to a few by using some eigenvalues of the covariance matrix (see e.g. [136]). Less known is the *Kernel Principal Component Analysis* (KPCA), which is aimed at overcoming the linearity of standard PCA [149] as well as other nonlinear techniques such as the method of principal curves [27]. Other nonlinear techniques for dimensionality reduction are special kinds of neural networks [16] and self-organizing maps [98]. Besides dimensionality reduction, a characteristic of these techniques for reliability analysis that can eventually be more important is the possibility of *problem transformation*, i.e. the conversion of the given problem into another one that can be more easily tractable for reliability analysis. For instance, the transformation of a reliability problem involving several limit-state functions (a series or a parallel problem) can be transformed into a problem with a single function by means of the nonlinear transformation implied by these techniques. This is demonstrated in this chapter with an example of KPCA.

For continuous random functions of time or space the analogous to PCA is the *Karhunen-Loéve transform*, which allows the representation of the function in terms of a few random variables [55]. From the viewpoint of Statistical Learning, this transform can also be considered an unsupervised learning method in that it extracts the structure of a random process. An alternative to this expansion is the wavelet transform of the function [108]. The wavelet transform has close relationships to the Karhunen-Loéve transform, as has been shown [181, 179].

Training Statistical Learning devices requires exploration of high dimensional spaces with a little number of samples. In some cases it is also desirable that the population be compressed without losing its random structure. Simple random generation of samples can be inefficient to such clustering because there is a high probability of uneven generation of the population. Data compression techniques such as *Vector Quantization* and *Expectation-Maximization* can be of avail in this respect, as they allow reducing a simulated population to the centers of clusters that are optimal in terms of encoding and decoding information. An exposition of these techniques closes the chapter.

3.2 Principal Component Analysis

This is a popular method for linear dimensionality reduction. The linearity of the method is due to the fact that the samples in the new (reduced) space are computed as linear combinations of the given ones. The method is also used for regression purposes.

Let $\boldsymbol{x}_i, i = 1, 2, \ldots, n$ denote a sample vector in $\mathbb{R}^d$ and assume that the vector is centered, i.e. the sample mean has been drawn. This implies that $\sum_{i=1}^{n} \boldsymbol{x}_i = \boldsymbol{0}$. An expansion of the form

$$\boldsymbol{x} = \sum_{j=1}^{d} v_j \boldsymbol{z}_j \tag{3.1}$$

is sought for an incoming vector $\boldsymbol{x}$, where v_j are coefficients and $\boldsymbol{z}_j$ basis vectors satisfying the orthonormality property

$$\langle \boldsymbol{z}_i, \boldsymbol{z}_j \rangle = \delta_{ij} \; . \tag{3.2}$$

Here $\langle \boldsymbol{a}, \boldsymbol{b} \rangle$ is the inner product in Euclidean space $\boldsymbol{a}^{\mathrm{T}} \boldsymbol{b}$ and δ_{ij} is the Delta of Kronecker:

$$\delta_{ij} = \begin{cases} 1, & i = j \\ 0, & i \neq j \; . \end{cases} \tag{3.3}$$

This condition implies that the coefficients v_j in the above expansion are

$$v_j = \langle \boldsymbol{z}_j, \boldsymbol{x} \rangle \; . \tag{3.4}$$

The expansion should be such that truncating this sum after $r < d$ terms minimizes the sum of squared errors. It can be demonstrated [16] that this can be achieved by diagonalizing the sample covariance matrix

$$\boldsymbol{C} = \frac{1}{n}\sum_{i=1}^{n} \boldsymbol{x}_i \otimes \boldsymbol{x}_i \ , \tag{3.5}$$

where $\otimes$ denotes the Kronecker product[1]. The diagonalization implies solving the eigenvalue problem

$$\lambda_k \boldsymbol{z}_k = \boldsymbol{C}\boldsymbol{z}_k \ , \tag{3.6}$$

where $\lambda_k, k = 1, 2, \ldots, d$ is an eigenvalue and $\boldsymbol{v}_k$ the corresponding eigenvector. From the numerical point of view, the eigenvalue problem is best solved out by means of the singular value decomposition of the matrix $\boldsymbol{X} = \{\boldsymbol{x}_i\}$. The d solutions are arranged according to the decreasing order of the eigenvalues, which happen to be equal to the variance of the projections v_j given by (3.4). Hence the sum of these variances equals the trace of the sample covariance matrix $\boldsymbol{C}$. Therefore, selecting the first $r < d$ eigenvalues is equivalent to choose the variables explaining a large part of the variance, and hence of the entropy, of the population [166].

The first-, second-, etc *principal components* are such projections. Since they have been ordered according to a decreasing degree of variance, the truncation of the expansion (3.1) after r terms implies a dimensionality reduction from $\boldsymbol{x} = \{x_1, x_2, \ldots, x_d\}$ to $\boldsymbol{u} = \{v_1, v_2, \ldots, v_r\}$ such that $r < d$.

For the next exposition of the Kernel PCA, it is important to make some further derivations [148]. Replacing (3.5) upon (3.6) yields

$$\lambda_k \boldsymbol{z}_k = \frac{1}{n}\sum_{i=1}^{n} (\boldsymbol{x}_i \otimes \boldsymbol{x}_i)\boldsymbol{z}_k \tag{3.7}$$

which, upon considering the properties of Kronecker product is equal to

$$\lambda_k \boldsymbol{z}_k = \frac{1}{n}\sum_{i=1}^{n} \langle \boldsymbol{x}_i, \boldsymbol{z}_k \rangle \boldsymbol{x}_i \ . \tag{3.8}$$

This equation indicates that all solutions $\boldsymbol{z}_k$ with nonzero eigenvalue λ_k belong to the span of the vectors $\boldsymbol{x}_i$ and hence can be written as

$$\boldsymbol{z}_k = \sum_{i=1}^{n} \gamma_{ik} \boldsymbol{x}_i \ . \tag{3.9}$$

Finally, multiplying (3.6) from the left by $\boldsymbol{x}_i^{\mathrm{T}}$ one obtains

$$\lambda_k \langle \boldsymbol{x}_i, \boldsymbol{z}_k \rangle = \langle \boldsymbol{x}_i, \boldsymbol{C}\boldsymbol{z}_k \rangle \ . \tag{3.10}$$

Principal Component Analysis can be used in time-invariant reliability problems having many mutually correlated variables for operating an effective

[1]Kronecker notation has been adopted herein. The Kronecker product $\boldsymbol{a} \otimes \boldsymbol{b}$ of vectors $\boldsymbol{a}$ and $\boldsymbol{b}$ is equivalent to $\boldsymbol{a}\boldsymbol{b}^{\mathrm{T}}$.

dimensionality reduction. Also, it has been used in structural reliability for the treatment of random processes [24].

3.3 Kernel PCA

Besides the methods of Principal Curves [27] and Neural PCA [37], Kernel PCA is a novel method aimed at the extension of standard PCA to nonlinear projection [148]. It differs from standard PCA in that the approach is based on a nonlinear transformation applied to the data before solving the eigenvalue problem. The method is based on the notion of kernels in a high-dimensional Hilbert space which is shared by the Support Vector Machines exposed in Chap. 5. The essentials of the kernel formulation will be exposed herein.

3.3.1 Basic Equations

Instead for searching a nonlinear relationship between the principal components and the original variables as in the Principal Curves method, the KPCA formulates this relationship in a linear form, thus preserving its mathematical advantages, after a previous nonlinear projection of the variables themselves on a space of higher dimensionality. Hence the (possibly nonlinear) mapping $\zeta : \mathbb{R}^d \mapsto \mathcal{Q}$ is applied to the given variables, where $\mathcal{Q}$ is called the *feature space*. It can be very large, virtually with infinite dimensions, as will be shown later. The nonlinear projection is, therefore $\boldsymbol{x} \mapsto \zeta(\boldsymbol{x})$. Assuming that the new variables are also centered around their mean, the covariance matrix in the new space is

$$\boldsymbol{C} = \frac{1}{n} \sum_{i=1}^{n} \zeta(\boldsymbol{x}_i) \otimes \zeta(\boldsymbol{x}_i) \tag{3.11}$$

according to (3.5). The task is again to solve the eigenproblem

$$\lambda_k \boldsymbol{z}_k = \boldsymbol{C} \boldsymbol{z}_k \tag{3.12}$$

for nonzero eigenvectors. The solution can be directly formulated in the form (3.10), i.e.

$$\lambda_k \big\langle \zeta(\boldsymbol{x}_i), \boldsymbol{z}_k \big\rangle = \big\langle \zeta(\boldsymbol{x}_i), \boldsymbol{C} \boldsymbol{z}_k \big\rangle \,, \tag{3.13}$$

where the eigenvectors satisfy

$$\boldsymbol{z}_k = \sum_{i=1}^{n} \gamma_{ik} \zeta(\boldsymbol{x}_i) \tag{3.14}$$

which is the analogous to (3.9). Substituting this equation into (3.13) yields

$$\lambda_k \sum_{i=1}^{n} \gamma_{ik} \langle \boldsymbol{\zeta}(\boldsymbol{x}_k), \boldsymbol{\zeta}(\boldsymbol{x}_i) \rangle = \frac{1}{n} \sum_{i=1}^{n} \gamma_{ik} \Big\langle \boldsymbol{\zeta}(\boldsymbol{x}_k), \sum_{j=1}^{n} \boldsymbol{\zeta}(\boldsymbol{x}_j) \langle \boldsymbol{\zeta}(\boldsymbol{x}_j), \boldsymbol{\zeta}(\boldsymbol{x}_i) \rangle \Big\rangle \quad (3.15)$$

which can be presented in a more compact form as

$$n\lambda_k \boldsymbol{K} \boldsymbol{b}_k = \boldsymbol{K}^2 \underline{b}_k \ , \quad (3.16)$$

where $\boldsymbol{K}_{ij} = \langle \boldsymbol{\zeta}(\boldsymbol{x}_i), \boldsymbol{\zeta}(\boldsymbol{x}_j) \rangle$ and $\boldsymbol{b}_k = [\gamma_{1k}, \ldots, \gamma_{nk}]^{\mathrm{T}}$. The solutions of this problem also satisfy

$$n\lambda_k \boldsymbol{b}_k = \boldsymbol{K} \boldsymbol{b}_k \quad (3.17)$$

which is more easily tractable. The eigenvalues are ordered in the same manner as in PCA. Upon imposing the normalization condition (3.2) one obtains after (3.14), the following:

$$\begin{aligned} 1 &= \sum_{i,j=1}^{n} \gamma_{ik}\gamma_{jk} \langle \boldsymbol{\zeta}(\boldsymbol{x}_i), \boldsymbol{\zeta}(\boldsymbol{x}_j) \rangle \\ &= \sum_{i,j=1}^{n} \gamma_{ik}\gamma_{jk} \boldsymbol{K}_{ij} \\ &= \langle \boldsymbol{\gamma}_k, \boldsymbol{K} \boldsymbol{\gamma}_k \rangle \\ &= \lambda_k \langle \boldsymbol{\gamma}_k, \boldsymbol{\gamma}_k \rangle \ . \end{aligned} \quad (3.18)$$

In the preceding developments, it has been supposed that the data in the feature space are also centered. Since this is not generally so, it is necessary to diagonalize a different matrix given by

$$\bar{\boldsymbol{K}}_{ij} = (\boldsymbol{K} - \boldsymbol{N}\boldsymbol{K} - \boldsymbol{K}\boldsymbol{N} + \boldsymbol{N}\boldsymbol{K}\boldsymbol{N})_{ij} \ , \quad (3.19)$$

where $\boldsymbol{N}_{ij} = 1/n, \forall i, j$ [148]. The principal components can in principle be computed as projections of the features $\boldsymbol{\zeta}(\boldsymbol{x})$ onto the eigenvectors:

$$v_k = \langle \boldsymbol{z}_k, \boldsymbol{\zeta}(\boldsymbol{x}) \rangle = \sum_{i,j=1}^{n} \gamma_{ik} \langle \boldsymbol{\zeta}(\boldsymbol{x}_i), \boldsymbol{\zeta}(\boldsymbol{x}) \rangle \ . \quad (3.20)$$

Note, however, that the nonlinear mapping appears in this equation in the form of an inner product, which is the operation defining the Hilbert space of functions. Hence, the principal components can be presented in the form

$$v_k = \langle \boldsymbol{z}_k, \boldsymbol{\zeta}(\boldsymbol{x}) \rangle = \sum_{i,j=1}^{n} \gamma_{ik} K(\boldsymbol{x}_i, \boldsymbol{x}) \ , \quad (3.21)$$

where the kernel $K(\cdot, \cdot)$ is defined as

$$K(\boldsymbol{x}, \boldsymbol{y}) = \langle \zeta(\boldsymbol{x}), \zeta(\boldsymbol{y}) \rangle \ . \quad (3.22)$$

Note that the conventional PCA is but a particular case of KPCA performed with the linear kernel $K(\boldsymbol{x}, \boldsymbol{y}) = \langle \boldsymbol{x}, \boldsymbol{y} \rangle$. The interesting point in the kernel formulation is that the nonlinear mapping $\zeta(\boldsymbol{x})$ need not be known explicitly. This is the so-called *kernel trick* in the literature on the subject [148]. It was first used in the theory of potential functions [1, 2] that constitutes the foundations of the Radial Basis Function Networks[2].

3.3.2 Kernel Properties and Construction

Given the importance of kernels, not only for KPCA but also for Support Vector Machines and derived learning methods, it is important to examine their properties and also the ways of building them. Let us first develop the expression for the so-called homogeneous polynomial kernel $K(\boldsymbol{x}, \boldsymbol{y}) = \langle \boldsymbol{x}, \boldsymbol{y} \rangle^p$. For instance, when $d = p = 2$ (two dimensions, second-degree kernel) we have

$$\begin{aligned} K(\boldsymbol{z}, \boldsymbol{y}) &= \left(\langle \boldsymbol{z}, \boldsymbol{y} \rangle\right)^2 \\ &= (z_1 y_1 + z_2 y_2)^2 \\ &= z_1^2 y_1^2 + z_2^2 y_2^2 + 2 z_1 y_1 z_2 y_2 \\ &= \langle \zeta(\boldsymbol{z}), \zeta(\boldsymbol{y}) \rangle \end{aligned} \tag{3.23}$$

where

$$\zeta(\boldsymbol{z}) = \{z_1^2, z_2^2, \sqrt{2} z_1 z_2\} \tag{3.24}$$

and similarly for $\zeta(\boldsymbol{y})$. The number of terms of the feature vector $\boldsymbol{\zeta}$ is

$$D = \binom{p + d - 1}{p} = \frac{(p + d - 1)!}{p!(d - 1)!} \tag{3.25}$$

which for $d = 5$ and $p = 2$ yields fifteen terms. For the inhomogeneous kernel $K(\boldsymbol{x}, \boldsymbol{y}) = \langle \boldsymbol{x}, \boldsymbol{y} + \theta \rangle^p$ the corresponding quantity is

$$D = \binom{p + d}{p} = \frac{(p + d)!}{p! d!} \tag{3.26}$$

which yields twenty-one terms for the same values of p and d. Hence, these nonlinear constructions represent a large increase of number of dimensions, which is a crucial fact exploited by support vector classification, as will be shown in Chaps. 5 and 7.

Another kernel that can be built starting from the nonlinear mapping is the Fourier kernel [178]. This way of constructing kernels, however, is very cumbersome and not optimal from the point of view of causing the just mentioned

[2]The name of potential functions derives from an electrostatic analogy used in the quoted seminal references. However, the expression of radial basis functions has been preferred in the statistical learning practice.

dimensionality explosion. The solution to this problem is given by *Mercer's theorem* for kernel functions in Hilbert spaces (see e.g. [137]). A Hilbert space $\mathcal{H}$ is a generalization of the Euclidean space $\mathcal{E}$ of vectors to the case of functions. The analogous to the inner product of vector in Euclidean space is the functional inner product

$$\langle f(t), h(t) \rangle = \int f(t)h(t)\mathrm{d}t \ . \tag{3.27}$$

Mercer's theorem states the following: For a function $f(\boldsymbol{x}) \in \mathcal{H}, \boldsymbol{x} \in \mathbb{R}^d$, such that its inner product with itself is finite,

$$\langle f(\boldsymbol{x}), f(\boldsymbol{x}) \rangle = \int f^2(\boldsymbol{x})\mathrm{d}\boldsymbol{x} < \infty \ , \tag{3.28}$$

and a symmetric kernel $K(\boldsymbol{x},\boldsymbol{y}) = K(\boldsymbol{y},\boldsymbol{x}), \boldsymbol{y} \in \mathbb{R}^d$ such that it is positive-definite,

$$\int K(\boldsymbol{x},\boldsymbol{y}) f(\boldsymbol{x}) f(\boldsymbol{y}) \mathrm{d}\boldsymbol{x}\mathrm{d}\boldsymbol{y} \geq 0 \ , \tag{3.29}$$

the kernel can be expanded in convergent series of the eigenfunctions $\psi(\cdot)$

$$K(\boldsymbol{x},\boldsymbol{y}) = \sum_{j=1}^{J} \lambda_j \psi_j(\boldsymbol{x}) \psi_j(\boldsymbol{y}) \tag{3.30}$$

corresponding to the J positive eigenvalues λ_j:

$$\lambda_j \psi_j(\boldsymbol{y}) = \int K(\boldsymbol{x},\boldsymbol{y}) \psi_j(\boldsymbol{x}) \mathrm{d}\boldsymbol{x} \ . \tag{3.31}$$

Applying the normalization $\zeta_j(\cdot) = \psi_j(\cdot)/\sqrt{\lambda_j}$ yields

$$\begin{aligned} K(\boldsymbol{x},\boldsymbol{y}) &= \sum_{j=1}^{J} \zeta_j(\boldsymbol{x}) \zeta_j(\boldsymbol{y}) \\ &= \langle \boldsymbol{\zeta}(\boldsymbol{x}), \boldsymbol{\zeta}(\boldsymbol{y}) \rangle \ . \end{aligned} \tag{3.32}$$

In words, a kernel satisfying the conditions for application of Mercer's theorem is the Euclidean inner product of vectors $\boldsymbol{\zeta}(\boldsymbol{x}), \boldsymbol{\zeta}(\boldsymbol{y})$, which remain implicit. Mercer's theorem assures that the implicit dimensionality of the feature space is virtually infinite, thus enormously facilitating hyperplane classification, as described in Chaps. 3 and 5. In practice, the positiveness of the kernel can be verified by the positive-definiteness of the associated matrix $\boldsymbol{K}$, obtained by its discretization over a square grid of points, as supported by the Moore-Aronzajn theorem [180].

It is easy to construct a Reproducing Kernel Hilbert Space (RKHS) $\mathcal{H}$ on functions on the input space, i.e. a space such that for a function

$$f(\boldsymbol{x}) = \sum_{j=1}^{\infty} \psi_j \zeta_j(\boldsymbol{x}) \tag{3.33}$$

its inner product with the kernel reproduces it:

$$\langle f(\boldsymbol{x}), K(\boldsymbol{x}, \boldsymbol{y}) \rangle_{\mathcal{H}} = f(\boldsymbol{y}) \, . \tag{3.34}$$

Here the inner product is defined as

$$\left\langle \sum_{j=1}^{\infty} \psi_j \phi_j(\boldsymbol{x}), \sum_{j=1}^{\infty} \psi'_j \phi_j(\boldsymbol{x}) \right\rangle_{\mathcal{H}} \equiv \sum_{j=1}^{\infty} \frac{\psi_j \psi'_j}{\lambda_j} \tag{3.35}$$

The most important consequence of construing kernels in this way is that the Hilbert space $\mathcal{H}$ does not contain functions that cannot be represented in the form given by (3.33) (see [33] for the details). Table 3.1 shows some kernels widely applied in kernel principal component analysis, support vector machine classification and other learning methods developed in the recent years [148]. The radial basis function and polynomial kernels are known to fulfill Mercer's conditions and for this reason they have found wide applicability in KPCA and SVM applications.

Table 3.1. Common kernels used by KPCA and SVM

Radial Basis Function	$K(\boldsymbol{x}, \boldsymbol{y}) = \exp\left(-\theta \lVert \boldsymbol{x} - \boldsymbol{y} \rVert^2\right)$
Polynomial	$K(\boldsymbol{x}, \boldsymbol{y}) = (\langle \boldsymbol{x}, \boldsymbol{y} \rangle + \theta)^p$
Sigmoidal	$K(\boldsymbol{x}, \boldsymbol{y}) = \tanh(\kappa \langle \boldsymbol{x}, \boldsymbol{y} \rangle + \theta)$
Inverse multiquadric	$K(\boldsymbol{x}, \boldsymbol{y}) = \frac{1}{[\lVert \boldsymbol{x} - \boldsymbol{y} \rVert^2 + \theta^2]}$

3.3.3 Example 1: Structure of a Monte Carlo Cloud

As a first illustration of the KPCA method for feature extraction, let us consider a cloud of random points discussed by [148]. It is shown in Fig. 3.1. The horizontal axis corresponds to a random variable $\underline{x}$ uniformly distributed in the range $[-1, +1]$ and the vertical one to $\underline{y} = \underline{x}^2 + \underline{n}$, where $\underline{n}$ is $\mathrm{N}(0, 0.2^2)$. A radial basis function kernel with $\theta = 10$ was used. The lines appearing in each panel correspond to contours of equal principal components for each eigenvalue. Since the cloud is symmetric in a statistical sense, there would be only two panels in conventional PCA, whose contours would be simple straight

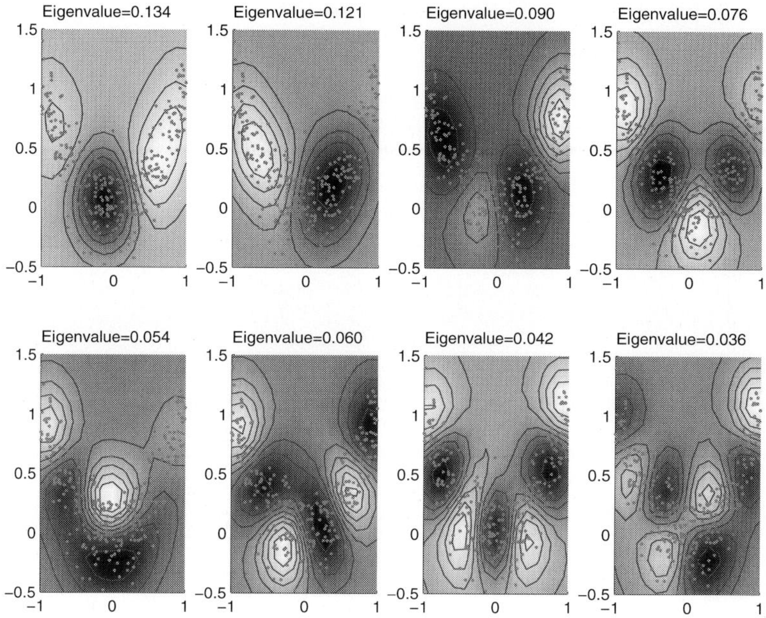

Fig. 3.1. Kernel Principal Component Analysis of a parabolic cloud of points

lines, vertical for the first eigenvalue and horizontal for the second. On the contrary, for the first few eigenvalues nonlinear PCA produces clusters whose contour lines are orthogonal to the cloud main axis at each portion, indicating the rotation of the central parabola. This makes the KPCA a promising method for analysing the results of a Monte Carlo simulation as in the tasks of model comparison and updating [41].

3.3.4 Example 2: Transformation of Reliability Problems

A second illustration concerns practical reliability problems treated under the classification paradigm (see Chaps. 1, 4 and 5). Figure 3.2 shows a parabolic limit-state function in standard Normal space and some safe and failure samples. A high curvature has been given to the function in order to induce a dispersion of the failure samples not only in the horizontal coordinate but also in the vertical one. This is done to represent a situation occurring in some reliability problems, where the unsafe samples tend to be widely spread along some coordinates due to their location in the tails of one or more random variables. As the number of dimensions increases this spread is much larger,

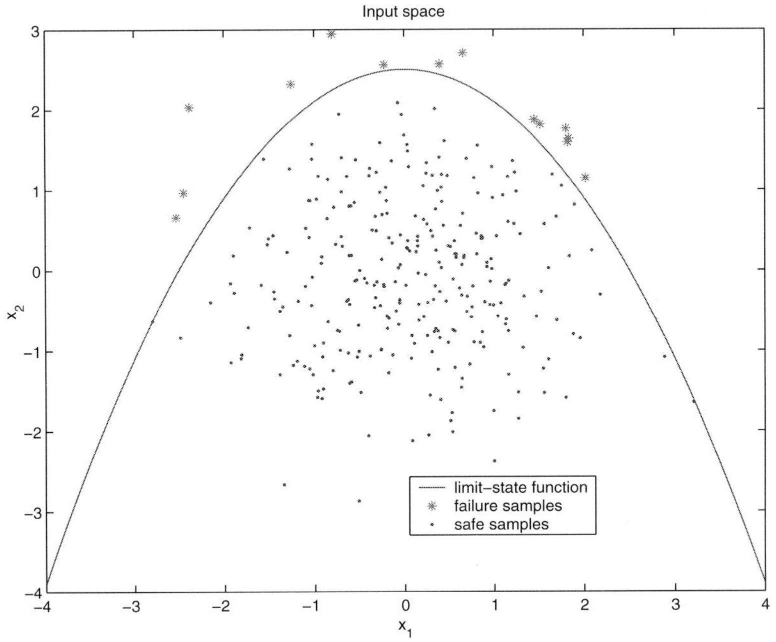

Fig. 3.2. KPCA applied to a reliability problem treated as a classification task. Notice that the unsafe samples are spread over a wide range in both coordinates

because this is the very expression of the curse of dimensionality (see Chap. 2). For training statistical learning devices this has the consequence that a very wide input space should be scanned in order to properly substitute the finite-element solver. Conventional PCA maintains such amplitude and, anyhow, the application of this method in standard $\boldsymbol{u}-$space is useless because it is constituted by uncorrelated variables. In this regard, Kernel Principal Component Analysis can be very useful as it can be used to produce a small cluster wherein all the few failure training samples are located, as illustrated by Fig. 3.3, where use was made of a RBF kernel. Similar clustering trends can also be seen in other relevant eigenvalue combinations. Using an adaptive-flexible classification method is easy to detect such clusters and therefore to reduce the domain wherein the decision function should be built.

Since KPCA allows clustering the failure samples in a small region in the nonlinear principal component space, it can be exploited for transforming a multiple reliability problem, like a series or parallel analysis, into a simple one. In order to illustrate this, consider Fig. 3.4 which shows two limit-state

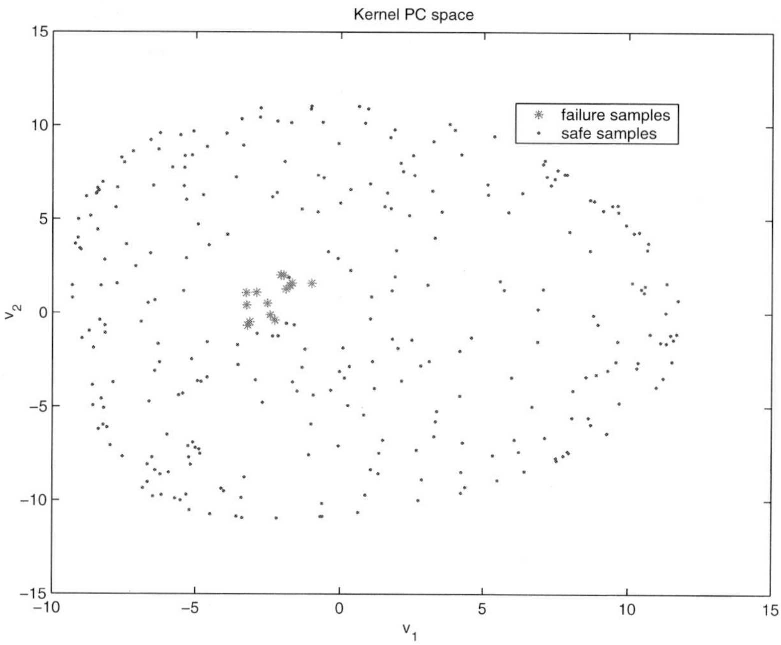

Fig. 3.3. KPCA applied to a reliability problem treated as a classification task. The plot corresponds to the first two nonlinear principal components. Notice that the failure samples have been clustered in a small domain that can be easily captured by a flexible classification technique

functions making a series problem and some failure and safe samples. These latter have been selected in a range close to the limit-state borders, as this suffices for training statistical learning devices of the supervised mode[3]. Next, Fig. 3.5 to 3.7 show the relationships between the first/second, first/third and second/third kernel principal components, respectively. In this calculation use was also made of the RBF kernel. It it evident that the multiple reliability problem has been effectively reduced to a single one characterized by a simply connected failure class that can be easily detected by an intelligent statistical device.

A word of caution, however, must be addressed at this point with respect to the practical use of the KPCA. Numerical experiments show that the results are highly dependent on the kernel and its parameters, so that this kind

[3]A general procedure for performing this selection is described in Chap. 4.

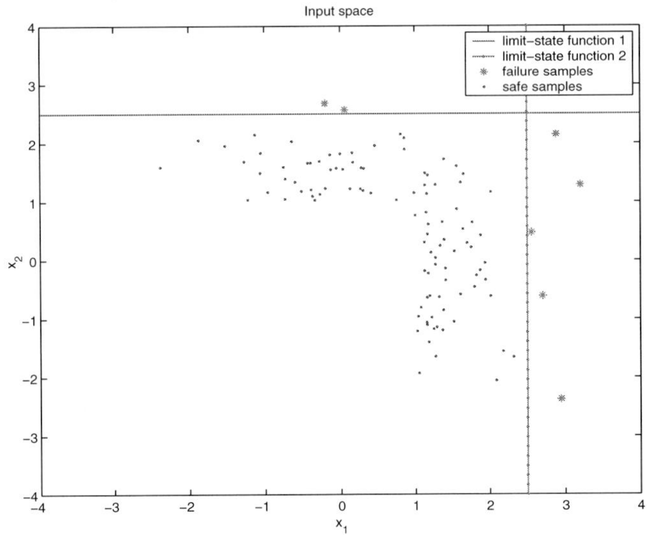

Fig. 3.4. KPCA applied to the reduction of a multiple reliability problem to a simple one. Problem definition in normalized input space

of preprocess cannot be applied in an automatic fashion without the participation of the analyst. Fortunately, the selection of the kernel can be facilitated by the graphical display of the relationships between the first few principal components. In general the application of this method in structural reliability applications is worth further research.

3.4 Karhunen-Loève Expansion

An important result of the theory of random process is the decomposition of the covariance function known as the *Karhunen-Loève decomposition* or *expansion*. In this paragraph we present its basic equations and properties, as it is a powerful device for dimensionality reduction in structural reliability, when it is necessary to model loads and material properties with random processes and fields.

Let us consider a zero-mean random process $\underline{x}(t)$. Since its covariance function $C(t_1, t_2)$ is positive definite (e.g. [158]), it has the spectral decomposition

$$C(t_1, t_2) = \sum_{i=1}^{\infty} \lambda_i f_i(t_1) f_i(t_2) \tag{3.36}$$

by virtue of Mercer's theorem 3.30. Here λ_i and $f_i(t)$ are the solutions of the eigenproblem

$$\lambda_i f_i(t_1) = \int C(t_1, t_2) f_i(t_2) \mathrm{d}t_2 \ . \tag{3.37}$$

The eigenfunctions are orthogonal:

$$\int f_i(t) f_j(t) \mathrm{d}t = 0, i \neq j \ . \tag{3.38}$$

For practical reasons they are usually normalized, i.e.

$$\int f_i(t) f_j(t) \mathrm{d}t = \delta_{ij} \tag{3.39}$$

where δ_{ij} is Kronecker's delta.

Let us now assume that a realization of a zero-mean random process $\underline{x}(t)$ can be expanded in the form

$$x(t) = \sum_{i=1}^{\infty} \sqrt{\lambda_i} f_i(t) u_i \tag{3.40}$$

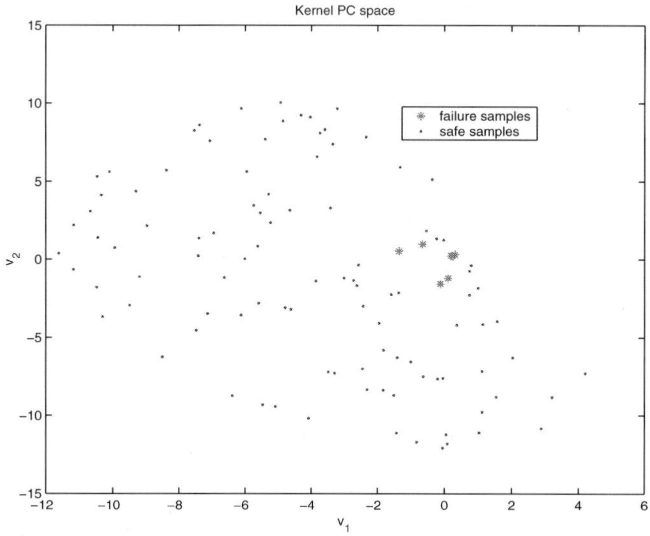

Fig. 3.5. Relationship between the first and second kernel principal components of samples in Fig. 3.4

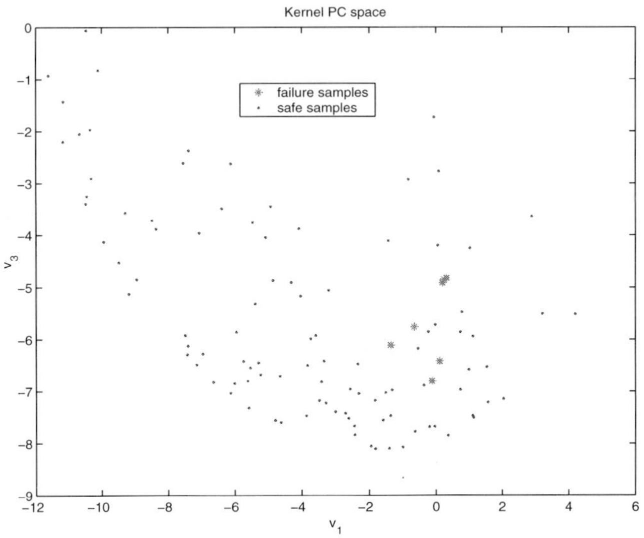

Fig. 3.6. Relationship between the first and third kernel principal components of samples in Fig. 3.4

where $\underline{u}_i$ are random variables. This implies that the realization is determined by summing up deterministic functions of time affected in amplitude by a constant and a random value whose properties are still to be determined. The covariance function of the realization becomes

$$C(t_1, t_2) = \mathrm{E}[\underline{x}(t_1)\underline{x}(t_2)] = \sum_{i=1}^{\infty}\sum_{j=1}^{\infty} \sqrt{\lambda_i}\sqrt{\lambda_j} f_i(t_1) f_j(t_2) \mathrm{E}[\underline{u}_i \underline{u}_j] \,. \tag{3.41}$$

Multiplying both sides by $f_k(t_2)$ and integrating over the whole real line yields

$$\begin{aligned} \int C(t_1, t_2) f_k(t_2) \mathrm{d}t_2 &= \int \sum_{i=1}^{\infty}\sum_{j=1}^{\infty} \sqrt{\lambda_i}\sqrt{\lambda_j} f_i(t_1) f_j(t_2) f_k(t_2) \mathrm{E}[\underline{u}_i \underline{u}_j] \mathrm{d}t_2 \\ &= \sum_{i=1}^{\infty}\sum_{j=1}^{\infty} \sqrt{\lambda_i}\sqrt{\lambda_j} f_i(t_1) \mathrm{E}[\underline{u}_i \underline{u}_j] \int f_j(t_2) f_k(t_2) \mathrm{d}t_2 \\ &= \sum_{i=1}^{\infty}\sum_{j=1}^{\infty} \sqrt{\lambda_i}\sqrt{\lambda_j} f_i(t_1) \mathrm{E}[\underline{u}_i \underline{u}_j] \delta_{jk} \end{aligned} \tag{3.42}$$

$$= \sum_{i=1}^{\infty} \sqrt{\lambda_i}\sqrt{\lambda_k} f_i(t_1) \mathrm{E}[\underline{u}_i \underline{u}_k] \, .$$

This means that

$$\lambda_k f_k(t_1) = \sqrt{\lambda_k} \sum_{i=1}^{\infty} \sqrt{\lambda_i} f_i(t_1) \mathrm{E}[\underline{u}_i \underline{u}_k] \, . \tag{3.43}$$

Multiplying this equation by $f_m(t_1)$ and integrating we get

$$\lambda_k \int f_k(t_1) f_m(t_1) \mathrm{d}t_1 = \sqrt{\lambda_k} \sum_{i=1}^{\infty} \sqrt{\lambda_i} \mathrm{E}[\underline{u}_i \underline{u}_k] \delta_{im} \, . \tag{3.44}$$

Considering (3.39) and (3.3) we obtain

$$\lambda_k \delta_{km} = \sqrt{\lambda_k \lambda_m} \mathrm{E}[\underline{u}_i \underline{u}_k] \tag{3.45}$$

which reduces to the simpler form

$$\mathrm{E}[\underline{u}_k \underline{u}_m] = \delta_{km} \, . \tag{3.46}$$

This indicates that the realizations of the zero-mean process given by (3.40) can be generated after a random variable $\underline{u}_i$ with properties

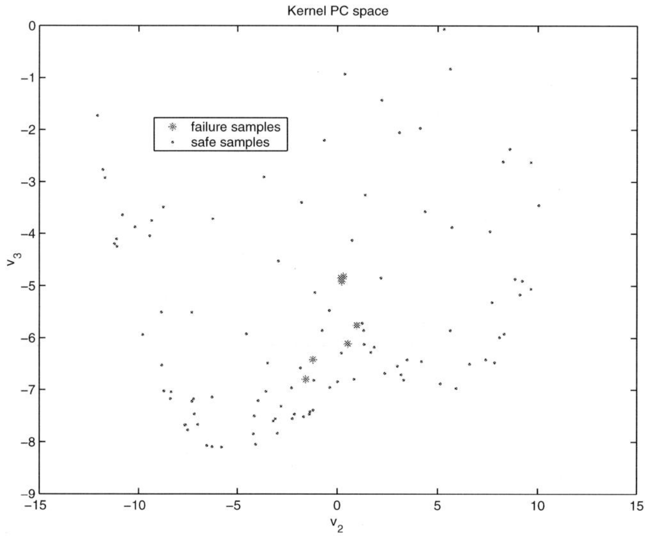

Fig. 3.7. Relationship between the second and third kernel principal components in Fig. 3.4

$$\mathrm{E}[\underline{u}_i] = 0 \qquad \mathrm{E}[\underline{u}_i \underline{u}_j] = \delta_{ij} \ . \tag{3.47}$$

Therefore, for simulating a Gaussian process it is sufficient to generate random variables $\underline{u}_i \sim \mathrm{N}(0,1)$ due to the linearity of the transformation (3.40). For solving the eigenvalue problem finite element approaches have been proposed [55, 72]. For problems not involving complex physical domains use can be made of the Nyström method [6], which requires much less computational effort [78]. A method based on wavelets was recently proposed by [125]. The expansion has been widely used in recent years for solving several stochastic mechanics problems [55, 56, 52, 54, 129, 146]).

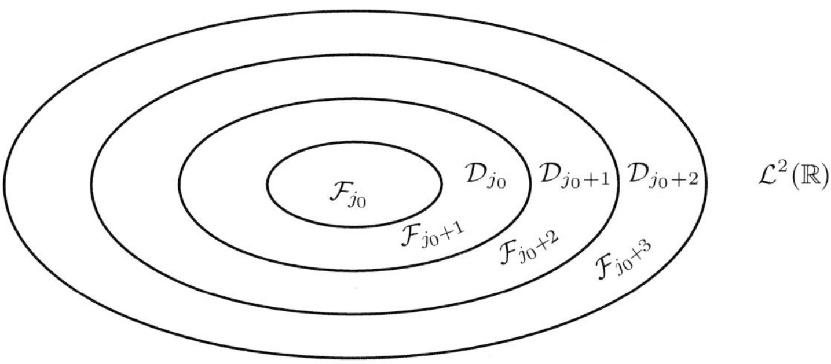

Fig. 3.8. Wavelet nested spaces

3.5 Discrete Wavelet Transform

Wavelet transforms have emerged in the last decade as a powerful method for signal analysis and processing [108]. They have distinguishing advantages over Fourier transforms for the analysis of nonstationary processes. One of the keys for the success of wavelets in such a field is their capability for signal compression and encryption, denoising and smoothing. All these tasks have in common the need for modifying the wavelet coefficients in such a way that the essential information is preserved. The transform allows such a modification at a high rate of essential information preservation. For this reason, it is also a very suitable method for extracting parameters from a random signal, since this task amounts to be the same as the data compression needed in signal transmission. In this respect the transform is especially useful when there is no knowledge of the covariance kernel of spatial or temporal processes, so that the Karhunen-Loève expansion cannot be readily applied. Such is the case, for instance, of input processes modeled as the result of complex physical phenomena or of the response processes of nonlinear structures.

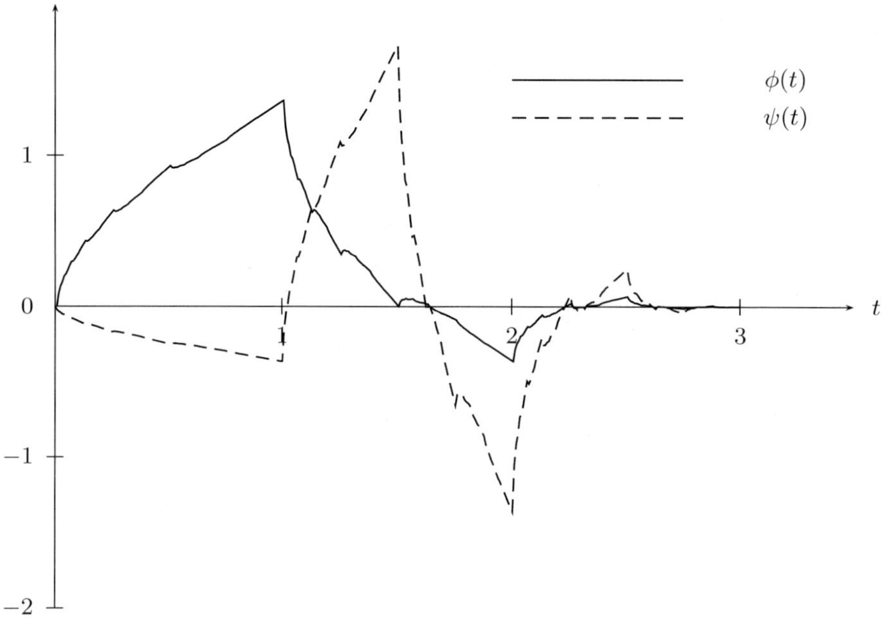

Fig. 3.9. Daubechies scaling and wavelet functions of order 4

The basic ingredient in the wavelet representation is a given *scaling function* $\phi(t) \in \mathcal{L}^2(\mathbb{R})$, for which several proposals have been developed (with the names of their developers, such as Haar, Daubechies, Meyer, etc.). A set of translations of this function is defined as $\phi_k(t) = \phi(t-k), k \in \mathbb{Z}$. Let us call $\mathcal{F}_0$ the subspace of $\mathcal{L}^2(\mathbb{R})$ spanned by functions $\phi_k(t)$. This means that any function in this space $f(t) \in \mathcal{F}_0$ can be represented as the linear combination

$$f(t) = \sum_k a_k \phi_k(t) , \tag{3.48}$$

where a_k are coefficients. The space $\mathcal{F}_0$, however, can be altered by means of two-dimensional basis functions defined as

$$\phi_{j,k}(t) = 2^{j/2}\phi(2^j t - k), \;\; j,k \in \mathbb{Z} , \tag{3.49}$$

where the term $2^{j/2}$ is a normalization coefficient. The effect of the factor 2^j in the parentheses is to produce a dilatation (for $j < 0$) or contraction (for $j > 0$) of the function in the time axis.

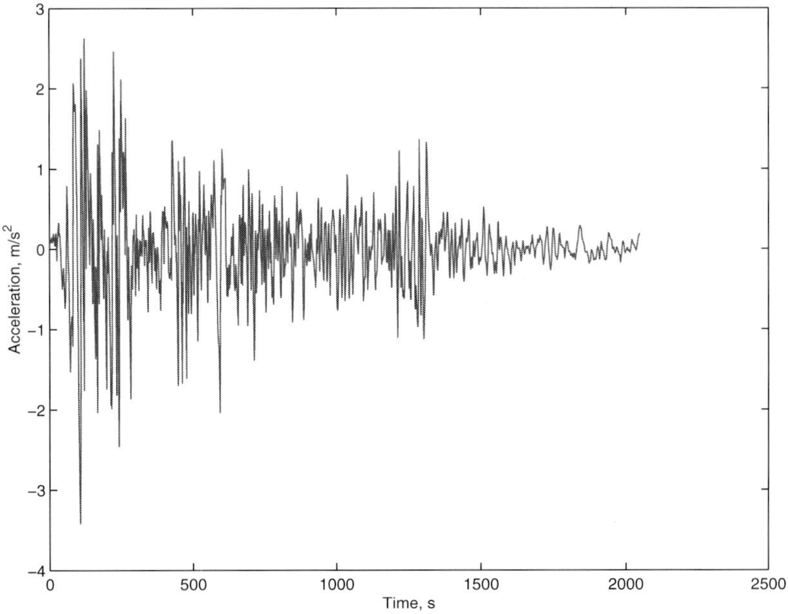

Fig. 3.10. El Centro (1949) earthquake record

Let $\mathcal{F}_j$ denote the space spanned by functions $\phi_{j,k}$. Hence we have a nested structure of spaces $\mathcal{F}_0 \subset \mathcal{F}_1 \subset \mathcal{F}_2 \ldots \subset \mathcal{L}^2(\mathbb{R})$. Notice the spaces have been defined such that the belonging of a function to the one space $f(t) \in \mathcal{F}_j$ implies and is implied by the belonging to the next one, $f(2t) \in \mathcal{F}_{j+1}$. Since the basis of each space are also functions contained in it, we have $\phi(t) \in \mathcal{F}_j$ and therefore $\phi(t) \in \mathcal{F}_{j+1}$. For this reason they can be expressed in terms of the basis of the next space:

$$\phi(t) = \sum_n a_n \phi(2t - n), \quad n \in \mathbb{Z} \, . \tag{3.50}$$

In this equation the coefficients a_n are commonly set equal to $\sqrt{2} r_n$ for normalization conditions. Part of the object of wavelet design is to find the necessary coefficients r_n.

Useful as this construction may be, it however lacks the powerful element of the wavelets as such, which are functions intended to span, not the space $\mathcal{F}_j$, but the difference space between two consecutive spaces. This difference space $\mathcal{D}_j$ is specified as the orthogonal complements of $\mathcal{F}_j$ in $\mathcal{F}_{j+1}$ (see Fig. 3.8):

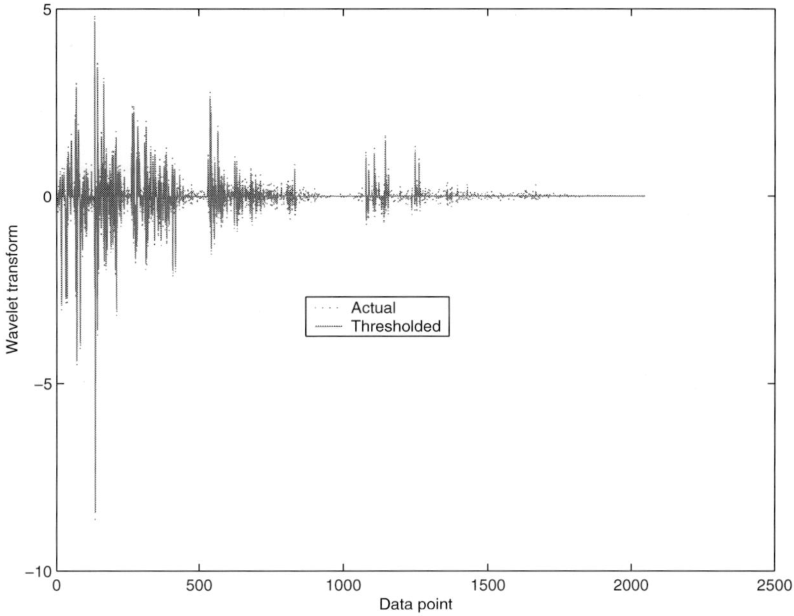

Fig. 3.11. Actual and thresholded inverse wavelet transform of El Centro record

$$\mathcal{F}_{j+1} = \mathcal{F}_j \bigoplus \mathcal{D}_j \ , \tag{3.51}$$

where the symbol $\bigoplus$ stands for the direct sum of spaces. The orthogonality is defined by requiring that the inner product of the basis of the two complementary spaces be null:

$$\langle \phi_{j,k}(t), \psi_{j,m}(t) \rangle = \int \phi_{j,k}(t)\psi_{j,m}(t)\mathrm{d}t = 0 \ , \tag{3.52}$$

The crucial point is that, similarly to (3.50), the basis function of the difference space can also be expressed in terms of the scaling function of the next space:

$$\psi(t) = \sum_n b_n \phi(2t - n), \ \ n \in \mathbb{Z} \ , \tag{3.53}$$

where $b_n = \sqrt{2} s_n$. This function can be used to produce dilatations and translations as in (3.49):

$$\psi_{j,k}(t) = 2^{j/2} \psi(2^j t - k) \ . \tag{3.54}$$

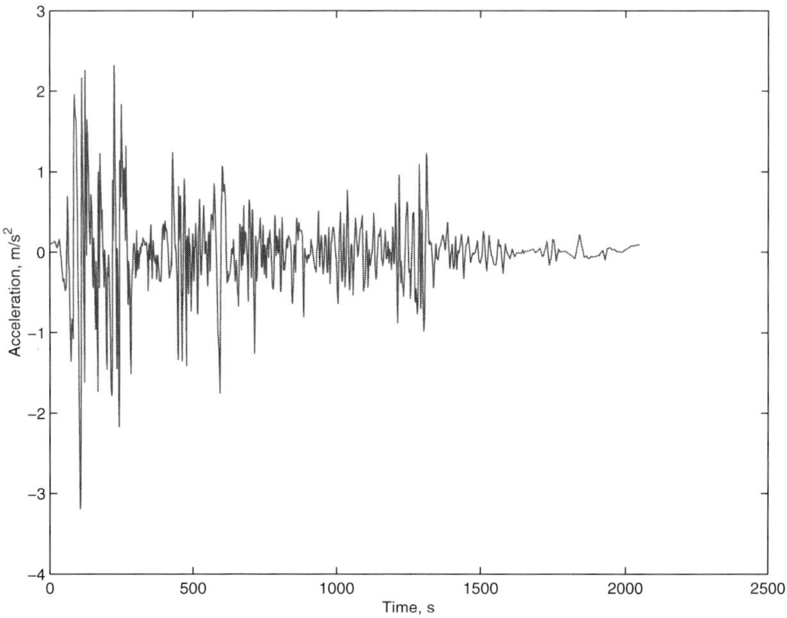

Fig. 3.12. Inverse wavelet transform of El Centro record after applying the shrinkage operation

Note, however, that since the wavelets are defined in terms of a so-called *mother wavelet* $\psi(t)$ and, in turn, this latter in terms of the scaling function, which is the basis of a certain space, the scale of the first $\phi_j(t)$ (defined by the coefficient j) is arbitrary. For this reason, we can say that

$$\mathcal{L}^2(\mathbb{R}) \equiv \mathcal{F}_{j_0} \bigoplus \mathcal{D}_{j_0} \bigoplus \mathcal{D}_{j_0+1} \bigoplus \mathcal{D}_{j_0+2} \bigoplus \cdots , \tag{3.55}$$

where j_0 denotes the starting resolution level. This end construction is depicted in Figure 3.8. Finally, it is important to note that the coefficients r_n and s_n are linked by the equation

$$s_n = (-1)^n r_{N-1-n} , \tag{3.56}$$

where N is the length of the signal vector. As an illustration, Fig. 3.9 shows the Daubechies wavelet and scaling function of order 4 (i.e., requiring four coefficients s_n and r_n for their definition). The end expression for the representation of the function $f(t)$ is, therefore,

$$f(t) = \sum_k \alpha_{j_0,k} 2^{j_0/2} \phi(2_0^j t - k) + \sum_k \sum_{j=j_0}^{\infty} \beta_{j,k} 2^{j/2} \psi(2^j t - k) \ , \tag{3.57}$$

where the coefficients α, β can be obtained by formulating the inner product of the target function $f(t)$ with the corresponding scaling or wavelet function. Taking into account the orthogonality conditions (3.52) the result is

$$\begin{aligned} \alpha_{j,k} &= \int f(t)\phi_{j,k}(t)\mathrm{d}t \ , \\ \beta_{j,k} &= \int f(t)\psi_{j,k}(t)\mathrm{d}t \ . \end{aligned} \tag{3.58}$$

Wavelet feature extraction can be accomplished by applying several shrinkage techniques. A popular one is the *soft thresholding* method [179], in which the new wavelet coefficients are defined

$$\tilde{\gamma} = \mathrm{sgn}(\gamma)(|\gamma| - \lambda)_+ \ , \tag{3.59}$$

where γ stands for either α or β, λ is a threshold and $(z)_+$ is the ramp function (i.e. zero for $z < 0$ and z for $z > 0$). This operation produces a new sequence of wavelet coefficients several of which are null. The compressed signal can be recovered by applying the inverse wavelet transform (3.57).

In order to illustrate the feature extraction capabilities of wavelets, the record of El Centro (1940) earthquake appearing in figure 3.10, which comprises 2048 data points, was used. The signal was transformed with Daubechies wavelets of order 6. A value of $\lambda = 0.2$ was adopted. As a result, only the 539 non-zero coefficients shown in Fig. 3.11 are needed to producing the signal appearing in Fig. 3.12, whose similarity with the original one is evident. The compression ratio is hence 3.8. If, in addition, the coefficients are hierarchically ordered and only a portion of the earthquake excitation matters for describing the structural response, the set of coefficient can be further reduced. This means that the selected wavelet coefficients can be sufficient for describing the input variables that have the strongest effect in the output response in a learning machine. Similarly to the Karhunen-Loève expansion, wavelets have been applied to the representation of random processes and fields [159, 186].

3.6 Data Compression Techniques

In training statistical learning devices in structural reliability problems it is often necessary to select a small subset of samples out of a large population. It was said in the introduction to this chapter that a simple Monte Carlo simulation is not efficient for this kind of preprocess because it is probable that the selection is not evenly distributed. *Data compression* (or *clustering*) techniques such as Vector Quantization or Expectation-Maximization can be

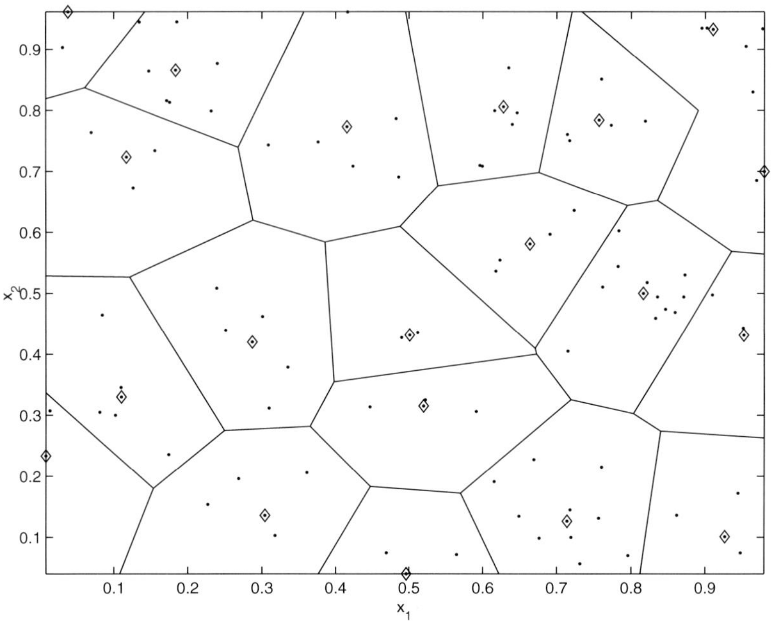

Fig. 3.13. Vector quantizers and Voronoi partition

of avail in this respect, as they allow reducing a simulated population to the centers of clusters that are optimal in terms of encoding and decoding information. These two techniques are described in this section.

3.6.1 Vector Quantization

A Vector Quantizer Q is a map $\mathbb{R}^d \mapsto \mathcal{R}$, where $\mathcal{R} = \{\boldsymbol{r}_1, \boldsymbol{r}_2, \ldots, \boldsymbol{r}_l\}$ is a finite subset of $\mathbb{R}^d$. In words, the l samples in $\mathcal{R}$ represent the n samples $\boldsymbol{x}_i, i = 1, 2, \ldots, n$. In order to obtain these samples the map must be associated to a partition $\pi_1, \pi_2, \ldots, \pi_l$ of $\mathbb{R}^d$ defined as the inverse operation to the quantization [27]:

$$\pi_k = Q^{-1}(\boldsymbol{r}_k) = \{\boldsymbol{x} \in \mathbb{R}^d : Q(\boldsymbol{x}) = \boldsymbol{r}_k\} . \tag{3.60}$$

This means that the quantizer performs the operations of encoding ($\mathbb{R}^d \mapsto \mathcal{R}$) and decoding ($\mathcal{R} \mapsto \mathbb{R}^d$) the samples. The quantizer is obtained when an optimal function

$$h(\boldsymbol{x},\boldsymbol{\omega}^*) = Q(\boldsymbol{x}) = \sum_{k=1}^{p} \boldsymbol{r}_k \Upsilon(\boldsymbol{x} \in \pi_k) \tag{3.61}$$

is found. The optimal character is given by a suitable measure of the distortion risk, i.e. the risk of false encoding, such as the mean quadratic risk given by

$$R(\boldsymbol{\omega}) = \int \|\boldsymbol{x} - h(\boldsymbol{x},\boldsymbol{\omega})\|^2 p_{\underline{\boldsymbol{x}}}(\boldsymbol{x})\mathrm{d}\boldsymbol{x} \; . \tag{3.62}$$

where $p_{\underline{\boldsymbol{x}}}(\boldsymbol{x})$ is the probability density function of the basic variables. The necessary conditions for calculating the quantizer upon this basis are known as *Lloyd conditions.* According to them, the optimal vector quantizers and the optimal partition are

$$\begin{aligned} \boldsymbol{r}_k &= \mathrm{E}[\boldsymbol{x}|\boldsymbol{x} \in \pi_k] \; , \\ \pi_k \supset \{\boldsymbol{x} \in \mathbb{R}^d &: \|\boldsymbol{x} - \boldsymbol{r}_k\| < \|\boldsymbol{x} - \boldsymbol{r}_j\|, \forall j \neq k\} \; . \end{aligned} \tag{3.63}$$

This partition is known in Spatial Statistics as the *Voronoi tessellation* [32] (see Fig. 3.13). In practice, the vector quantizers calculated over a population of distance-based selected samples can be used in high dimensional spaces to postulate a trial classifier. This is due to their ability to perform a dimensionality reduction, which have been exploited in several applications [98].

3.6.2 Expectation-Maximization

As said in Chap. 2, this method is widely applied for estimating multivariate density functions using linear combinations of Gaussian density functions. However, it can be used also for data compression, because the centers of the Gaussian functions play the role of the vector quantizers introduced in the previous section.

Consider the following estimate for the multivariate density underlying some samples of the vector random variable $\underline{\boldsymbol{x}}$ [34]:

$$p_{\underline{\boldsymbol{x}}}(\boldsymbol{x}) = \sum_{j=1}^{m} w_j \phi(\boldsymbol{x}; \boldsymbol{\mu}_j, \sigma_j) \; , \tag{3.64}$$

where w_j are parameters and $\phi(\boldsymbol{x}; \boldsymbol{\mu}_j, \sigma_j)$ is a multivariate Gaussian density function with mean $\boldsymbol{\mu}_j$ and covariance matrix $\sigma_j \boldsymbol{I}$. The above equation can be given a Bayesian interpretation in the form

$$p_{\underline{\boldsymbol{x}}}(\boldsymbol{x}) = \sum_{j}^{m} P(j) \phi(\boldsymbol{x}; \boldsymbol{\mu}_j, \sigma_j) \; , \tag{3.65}$$

in which subindex j is treated as a component having a certain probability of occurrence. Notice that the conditions

$$0 \leq P(j) \leq 1 \,,$$
$$\sum_j P(j) = 1 \tag{3.66}$$

must hold. On the basis of a given set of samples $\boldsymbol{X} = \{\boldsymbol{x}_i\}, i = 1, 2, \ldots, n$, the solution can be sought by applying the Maximum Likelihood Principle, which implies maximizing the function

$$L(\boldsymbol{X}) = \sum_{i=1}^{n} p_{\underline{\boldsymbol{x}}}(\boldsymbol{x}_i) \,. \tag{3.67}$$

This is equivalent to minimizing the empirical risk

$$R_{\mathrm{e}} = -\sum_{i=1}^{n} \ln \left(\sum_{j=1}^{m} P(j)\phi(\boldsymbol{x}_i; \boldsymbol{\mu}_j, \sigma_j) \right) \,. \tag{3.68}$$

The Expectation-Maximization algorithm is a procedure for calculating iteratively the quantities defining the model by treating it as a Bayesian classification problem in which the class label are not known in advance (this is called an *incomplete problem*) in pattern recognition. This is justified because in practice model (3.65) is used for generating samples $\boldsymbol{x}_i$ using the Gaussian component j with probability $P(j)$. Since these latter are constrained as indicated by (3.66), some auxiliary variables are introduced for obtaining an unconstrained optimization problem. The name of the procedure derives from the fact that maximizing the expected value of the log-likelihood function with respect to the auxiliary variables is equivalent to maximizing it with respect to the actual parameters. The result is the following procedure for updating the model parameters from the iteration step k to step $k+1$:

$$P^{[k+1]}(j|\boldsymbol{x}_i) = \frac{P^{[k]}(j)\phi(\boldsymbol{x}_i; \boldsymbol{\mu}_j[k], \sigma_j[k])}{\sum\limits_{i=1}^{n} P^{[k]}(j)\phi(\boldsymbol{x}_i; \boldsymbol{\mu}_j[k], \sigma_j[k])} \,,$$

$$\boldsymbol{\mu}_j[k+1] = \frac{\sum\limits_{i=1}^{n} P^{[k]}(j|\boldsymbol{x}_i)\boldsymbol{x}_i}{\sum\limits_{i=1}^{n} P^{[k]}(j|\boldsymbol{x}_i)} \,, \tag{3.69}$$

$$\sigma_j^2[k+1] = \frac{1}{d} \frac{\sum\limits_{i=1}^{n} P^{[k]}(j|\boldsymbol{x}_i)\|\boldsymbol{x}_i - \boldsymbol{\mu}_j[k+1]\|^2}{\sum\limits_{i=1}^{n} P^{[k]}(j|\boldsymbol{x}_i)} \,,$$

$$P^{[k+1]}(j) = \frac{1}{n} \sum_{i=1}^{n} P^{[k]}(j|\boldsymbol{x}_i) .$$

The procedure is started with reasonable trial values. See [111] for a detailed exposition of the method.

4

Classification Methods I - Neural Networks

4.1 Introduction

As noted in Chap. 1, the reliability problem can be regarded under a classification viewpoint, because the limit-state function determines two well defined classes of samples, namely the safe and failure ones. Despite this clear association, the structural reliability community has paid little attention to solve this problem under such a paradigm. This is perhaps due to the fact that classical pattern recognition methods are dominated by the Bayesian approach to class discrimination, whose application in structural reliability is not adequate as shown later on. The development of computers, however, has fostered the development of new statistical methods that are characterized by an abandonment of the parametric approach that dominated Statistics in the most part of the twentieth century and the use of either non-parametric or adaptive-flexible models. Nowadays, there are many pattern recognition methods that could be investigated with regard to their applicability in structural reliability (See [69, 42] and [148] for a detailed description of these methods). In order to facilitate this task, it is valuable to group them in some classes, according to their treatment of the training samples. The following classification is made herein with regard to their applicability to structural reliability analysis:

1. *Probabilistic methods*, which are based on the estimation of the density function or other statistical measures of the classes. Besides the classical Bayesian method, this group includes variants of the Radial Basis Function concept, because they have a clear Bayesian interpretation [16]. This comprehends the neural networks with that name as well as the Probabilistic Neural Networks, the Method of Potential Functions, etc. Also Linear Discriminant Analysis (LDA) and Logistic Regression belong to this group.
2. *Boundary methods.* This group comprehends methods that are aimed at the approximation of the boundary between the classes without regard to their statistical characterization. Among them are Rosenblatt's Percep-

tron, Multi-Layer Perceptrons, Support Vector Machines and their variants.
3. *Euclidean methods* (for want of a better name) in which the randomness of the training samples is not as decisive as their position in space. In this family we have the Nearest Neighbors, Classification Trees and others.

It is beyond the present monograph to examine all these methods in detail, not only for space reasons but more importantly because only a few of them are useful to the purpose at hand. In fact, a reading of the criteria for choosing learning methods in Structural Reliability listed in Sect. 2.12 indicates that only those methods that (a) could model a perfect, though nonlinear class separability without large errors and (b) that can be trained with a reduced number of samples are useful to this purpose. The revision of the above cited methods according to these criteria suggests that boundary methods such as Multi-Layer Perceptrons (MLP) and Support Vector Machines (SVM) are attractive for solving structural reliability problems regarded as pattern recognition tasks. This is because they can be trained with small sets of samples located in the vicinity of the limit-state function as they are addressed to the direct detection of such a boundary. However, since one of the purposes of this monograph is to discuss the classification approach at some length, due to the little attention that has been addressed to it in structural mechanics, it is important to show also why some representatives of the other two groups are beyond our interest. To this end, the classical Bayesian approach will be first discussed. It is demonstrated that its orientation to classification on the basis of probability density functions makes it very costly for structural reliability applications, in spite of the a priori knowledge of the joint density function of the basic variables. Another method of this group, namely Linear Discriminant Analysis also known as Fisher Discrimination, is examined in the section devoted to Rosenblatt's Perceptron, as this comparison is instructive on the difference between probabilistic and boundary methods and it serves as an introduction to the development of a secant reliability index made in Chap. 7.

Classification Trees, as a representative of the third group is then discussed. It is shown that despite this is a non-parametric method (in the usual sense of this word in Statistics, i.e. distribution-free) the rigidity of the model it implements makes it more convenient for natural problems showing non-perfect class separability but not for reliability problems in which the opposite situation occurs.

The chapter follows with a more lengthy examination of Multi-Layer Perceptrons, since these are a technique that can be applied either under the classification or regression paradigms. The latter application is discussed in Chap. 6. The description of the MLP is initiated with a description of hyperplane separation and Rosenblatt's Perceptron Algorithm, as these themes constitute a common basis for both the Multi-Layer Perceptrons and the Support Vector

Machines, which are the subject of the next chapter. Some examples illustrate their accuracy and economy when applied to reliability problems.

The chapter is closed with a similar description to Radial Basis Functions Networks (RBFN) and a discussion on their differences to Multi-Layer Perceptrons. The classification using RBFN and its variants is examined. It is shown that they exhibit the same problems of Bayesian classification, because they can be interpreted as a adaptive-flexible implementation of that classical approach. However, RBFN are useful for structural reliability analysis under a regression viewpoint, both as a method for estimating the failure probability as well as the entire distribution function of the observed responses. For this reason they are recovered in Chap. 6.

Before entering into the description of the categories of methods it is important to say that for normal pattern recognition problems dealing with natural samples belonging to non-perfectly separable classes, there is a so-called *no-free-lunch theorem* stating that, in the absence of assumptions, there is no motive to prefer a classification algorithm over another [42]. In words, there is need of empirical testing of many methods in order to select the best in each specific situation. In the structural reliability case, however, the specific features of its classification problem formulated in Sect. 2.12 play the role of such assumptions, thus implying that at least a class of methods should be evidently preferable over other classes and that no extensive empirical tests are necessary.

4.2 Probabilistic and Euclidean methods

Some classical discrimination methods are examined next with regard to their applicability for structural reliability analysis.

4.2.1 Bayesian Classification

This method is based on the well-known Bayes formula for updating conditional probabilities of random events. In this case these are the belonging of a sample to one class or to another. According to this rule, the vector $\boldsymbol{x}$ is assigned to the class $\mathcal{C}_k$ presenting the highest posterior probability $P(\mathcal{C}_k|\boldsymbol{x})$,

$$P(\mathcal{C}_k|\boldsymbol{x}) = \frac{p(\boldsymbol{x}|\mathcal{C}_k)P(\mathcal{C}_k)}{p(\boldsymbol{x})} \,. \tag{4.1}$$

This is equivalent to favor class k against class j if [139]

$$p(\boldsymbol{x}|\mathcal{C}_k)P(\mathcal{C}_k) > p(\boldsymbol{x}|\mathcal{C}_j)P(\mathcal{C}_j) \,. \tag{4.2}$$

For the two-class problem, this rule can be formulated as a function

$$g(\boldsymbol{x}) = \ln \frac{p(\boldsymbol{x}|\mathcal{C}_1)}{p(\boldsymbol{x}|\mathcal{C}_2)} + \ln \frac{P(\mathcal{C}_1)}{P(\mathcal{C}_2)} \tag{4.3}$$

such that the sign of $g(\boldsymbol{x})$ suffices for the classification: If $g(\boldsymbol{x}) > 0, \boldsymbol{x} \in \mathcal{C}_1$ and if $g(\boldsymbol{x}) < 0, \boldsymbol{x} \in \mathcal{C}_2$.

The prior probabilities $P(\mathcal{C}_k), k = 1, 2$ are calculated on the basis of training samples located in either class. The class-conditional densities $p(\boldsymbol{x}|\mathcal{C}_k)$ are commonly assumed as multivariate Gaussians using the mean vector and covariance matrix of the samples in each class or estimated using kernel or finite mixture models (see Chap. 2).

In the reliability analysis context, there is the paradox that the joint density function of the variables $p(\boldsymbol{x})$ is known, but this prior information cannot be incorporated in the calculation because it is a common factor to all classes and for this reason it does not appear in (4.2). Besides, the calculation of the class-conditional densities on this basis depends on the target of the calculation, namely the failure probability. In fact, taking a one-dimensional problem for simplicity, the following equations hold:

$$\begin{aligned} p(x|\mathcal{S}) &= \frac{p(x)}{P_{\mathrm{f}}}, \quad x \in \mathcal{S} \,, \\ p(x|\mathcal{F}) &= \frac{p(x)}{1 - P_{\mathrm{f}}}, \quad x \in \mathcal{F} \,. \end{aligned} \tag{4.4}$$

Such dependency makes impossible the direct use of the exact class-conditional densities. Since the probability of failure enters in these equations as a multiplicative constant, the class conditional densities could eventually be determined by any of the methods of the Metropolis family of algorithms [140], in which the constant cancels out. However, it is necessary to check whether each sample is in the safe or in the failure zone, as stated by the conditions appended to the equations. In other words, the samples must be classified in order to build the classifier, which amounts to a crude Monte Carlo simulation. This all means that the knowledge of the actual joint density function of the variables is of little use for Bayesian classification in this context.

A second drawback of the method is its high probability of error, which is given by the shaded area in Fig. 4.1. If the two classes are well apart, the area decreases and the confidence on the method increases. Unfortunately, in reliability analysis the two classes are joint to each other, thus making the application of the method subject to high uncertainties. Finally, a third problem is that the decision rule is clearly dependent on the training population, as evidenced by equation (4.3). In fact, the class-conditional densities and the prior probabilities differ if the training samples are generated after the actual joint density or, for instance, around the limit state function. This confirms that the method requires a classified training population that should resemble that with which the classifier will be used for calculating the failure probability. As a conclusion, the Bayesian method is suitable for working with natural

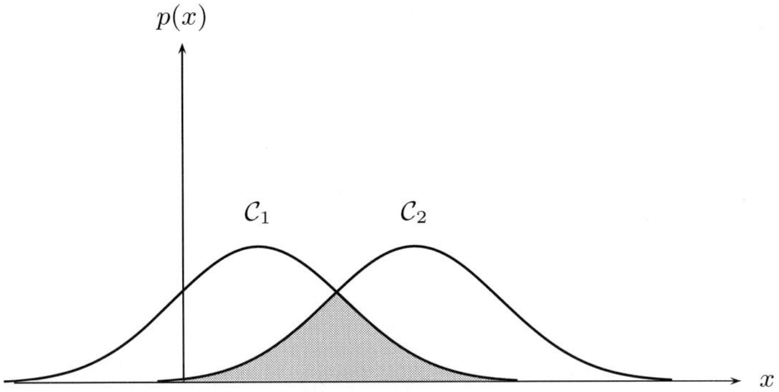

Fig. 4.1. Error probability in Bayesian classification

populations, in which training and test sets are expected to be similar, but not for working with reduced synthetic populations, such as those desired in structural reliability.

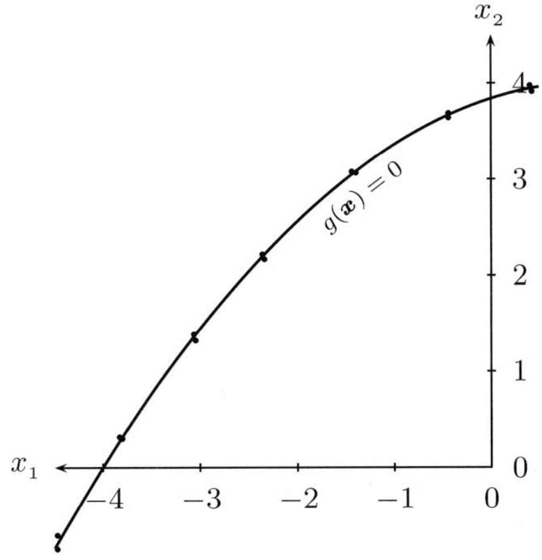

Fig. 4.2. Limit-state function and training set for testing classification methods

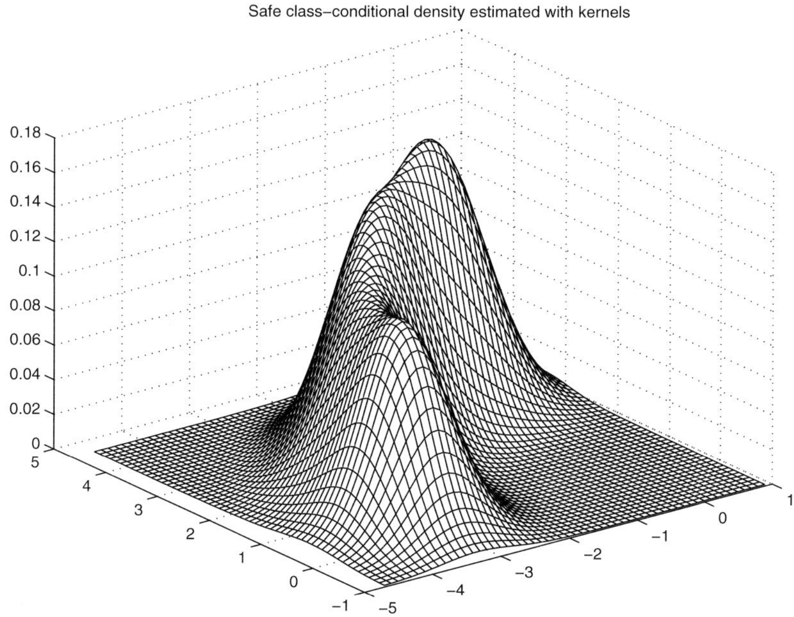

Fig. 4.3. Class density in the safe domain for Bayesian analysis

In order to illustrate these features the following limit state function [113] is used:

$$g(x_1, x_2) = 4 - 4(x_1 - 1)^2/25 - x_2 = 0 \;, \tag{4.5}$$

where $x_i, i = 1, 2$ are standard Normal variables. The function is shown in Fig. 4.2. Also shown in the figure are fourteen samples selected in the vicinity of the actual limit state function, which will be used for testing Classification Trees, Multi-Layer Perceptrons and Support Vector Machines in the sequel. Such a reduced sample set would be ideal for building a solver-surrogate, as pointed out by [96]. It is evident that upon this reduced sample set it is non-sense to build a Bayesian classifier, even with the aid of kernel density models. Therefore, a population of 180 samples embracing similarly the limit-state function was generated (see Fig. 5.8). The class densities in the safe and failure domains were estimated using the kernel model and the Expectation-Maximization algorithm summarized in Chap. 2. These densities appear in figures 4.3 and 4.4, respectively. On this basis and the a priori information $P(\mathcal{S}) = 100/180$ and $P(\mathcal{F}) = 80/180$ a failure probability estimate of 0.0084

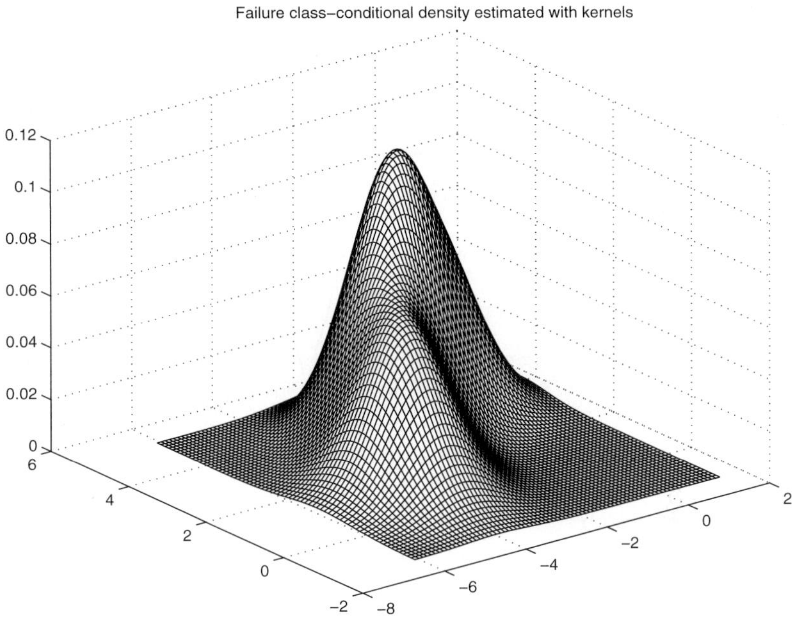

Fig. 4.4. Class density in the failure domain for Bayesian analysis

was obtained, which is very different from the actual one, 9.5×10^{-4}, which was obtained with 100,000 Monte Carlo samples.

This discussion of the Bayesian classification method illustrates the validity of the Vapnik's advice quoted often before about the avoidance of solving complicated intermediate problems (such as the class-conditional density estimation) in the way to solving simpler ones (i.e the estimation of a single probability). For all these reasons the method is not recommended to reliability analysis purposes. It is instead very suitable for classification of the abundant samples that can be drawn from experience with natural phenomena, which normally are not perfectly separable and hence the amount of error shown in Fig. 4.1 compares well to the amount of noise present in the training population.

4.2.2 Classification Trees

This is a classification technique developed in the eighties [20]. The samples are classified by checking inequalities of the form $x_i < \xi_i$, where $x_i, i = 1, 2, \ldots, d$ is a coordinate of the sample and ξ_i a reference value (see Fig. 4.5). If the in-

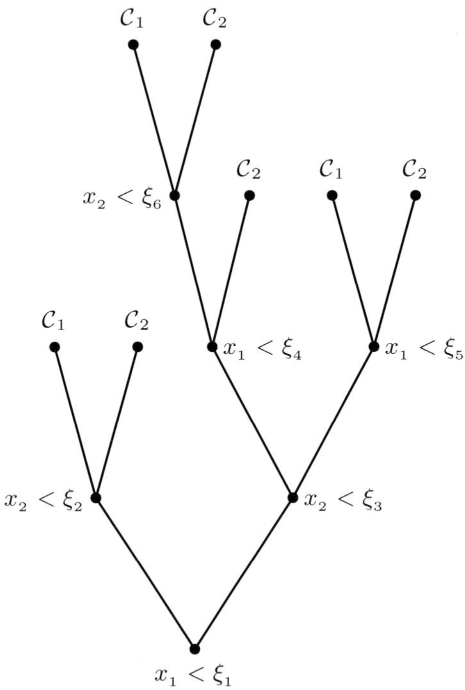

Fig. 4.5. A classification tree

equality holds then the left path must be followed; otherwise the classification process continues by right branch. At the end one arrives at the assigning of the sample to a class $\mathcal{C}_k$.

In order to obtain the values ξ_i it is necessary to apply a training method, whose essential element is the so-called *Impurity Function*, defined as the degree of mixing of samples belonging to different classes in a node t. One of the ways of measuring it is by means of the *Gini diversity index*, given by

$$i(t) = \sum_j \sum_{k \neq j} P(\mathcal{C}_k|t)P(\mathcal{C}_j|t) , \tag{4.6}$$

where $P(\mathcal{C}_k|t)$ is the probability that a sample corresponds to class k given that it has reached node t. The purpose of the training is to partition the

$d-$dimensional sample space into small boxes so that they are as pure as possible. At this point a regard should be paid to the theory of induction principles summarized in Chap. 2, because if the tree is trained so that all the boxes are pure the empirical risk will be null at the expense of elevating the risk of bad generalization. Consequently, it is necessary to apply a regularization approach of the form

$$R_{\mathrm{r}}(A) = R_{\mathrm{e}}(A) + \lambda \,\mathrm{card}(\bar{A}) \;, \tag{4.7}$$

where $R_{\mathrm{e}}(A)$ is the empirical risk of tree A, $\bar{A}$ is the set of all terminal nodes and $\mathrm{card}(\bar{A})$ is the cardinality of this set. It is evident that the second term of the right hand side of (4.7) defines the model complexity and that the parameter λ penalizes it. A little λ means that a complex tree is tolerated. On the contrary, a large value of λ leads to a tree with a few branches. In practice, the building of the tree consists in deciding the maximum number of samples in a box, then training the tree and finally pruning it. The first training yields a tree A_{max} while pruning the tree yield a sequence $k = 1, 2, \ldots, K$ of sub-trees which are more and more chopped. This can be expressed symbolically in the form

$$A_{\mathrm{max}} \supset A_1 \supset A_2 \ldots A_K = \{t_1\} \;, \tag{4.8}$$

where $A_K = \{t_1\}$ is a tree having only the initial node. To this sequence there correspond parameters

$$\lambda_1 < \lambda_2 < \ldots < \lambda_K \;. \tag{4.9}$$

It is observed that this structure has a flavor of the SRM principle, as this makes use of nested functional spaces (see Chap. 2). However, the nesting criterion is herein based on the regularization parameter instead of the VC dimension, which is a property of the set of classification functions.

This method was also tested with the limit-state function (4.5) using the fourteen points samples shown in Figures 4.2. The result is shown in Fig. 4.6. The dashed line is an interpretation of the inequalities defining the tree as the estimate of the limit state function. It can be seen that for such a small training population the tree is not useful as a solver surrogate since large errors can be expected. However, for a larger training set there is a remarkable increase of the accuracy. In fact, the tree was trained again with 180 training samples also lying in the surroundings of the function used previously for training the Bayes classifier. The result was 9.5×10^{-4}, which equals the exact result quoted above. This indicates that a good performance of the method for structural reliability analysis requires large sample sets. Despite being an adaptive method, this shortcoming is evidently a consequence of its rigidity, as it uses hyperplanes parallel to the axes of basic classifiers.

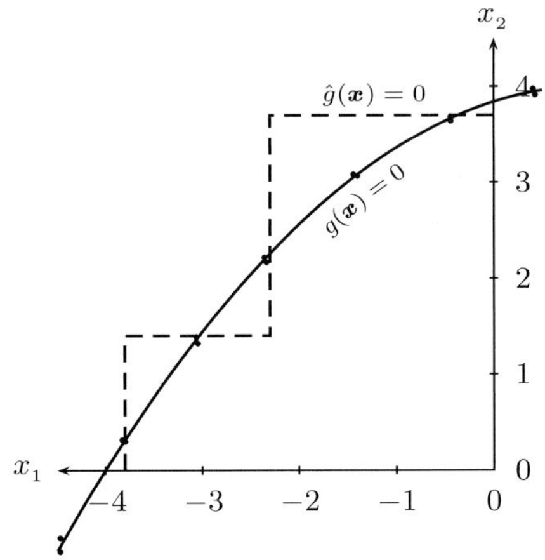

Fig. 4.6. Training set and limit state function obtained with a classification tree

4.2.3 Concluding Remarks

It has been shown in this section that two representative methods of the probabilistic and Euclidean groups are not convenient for structural reliability applications. However, this conclusion should not be generalized to all methods of the groups, because the categories are not so rigid and the methods in them are very different to each other. What is important in this regard is that methods showing the following characteristics are to be entirely avoided:

- Those requiring multivariate density estimation.
- Those implementing rigid models.

In this sense, it must be said that Nearest Neighbors, Constrained Topological Mapping and other methods that do not have these characteristics are worth testing in this field. Besides, in spite of Multi-Layer Perceptrons with hard-limiter activation functions implement a rigid model, they are examined in the next section not only because they are useful for understanding the operations performed by other MLP but also because they illustrate that the adaptivity of a learning device is not sufficient for structural reliability purposes if it is not accompanied by the flexibility characteristic.

4.3 Multi-Layer Perceptrons

Artificial neural networks are numerical algorithms that have their inspiration in the behavior of some biological structures, especially those found in the brain. An actual neuron is composed by a soma, some dendrites and an axon which ends in many branches that connect the neuron to other ones. The roles played by these components are as follows: The dendrites carry on the electrical impulses sent to the neuron by neighboring ones; the soma collects this information and the axon transforms it into a signal, which is sent to other neurons through the terminals. Many types of artificial neurons and networks have been proposed to imitate the behaviour of these biological cells. In this section we will introduce some general concepts on the neural network methodology and their application to pattern recognition. The issues examined are (a) the hyperplane separation, which also serves as an introduction to Support Vector Machines (SVM) dealt with in the next chapter and (b) the construction of Multi-Layer Perceptrons with hard-limiter activation functions.

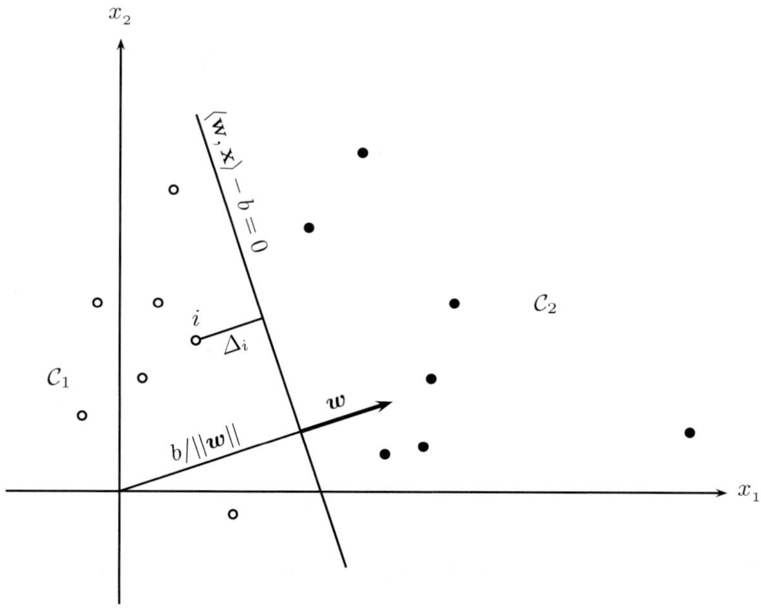

Fig. 4.7. Basic elements of hyperplane separation

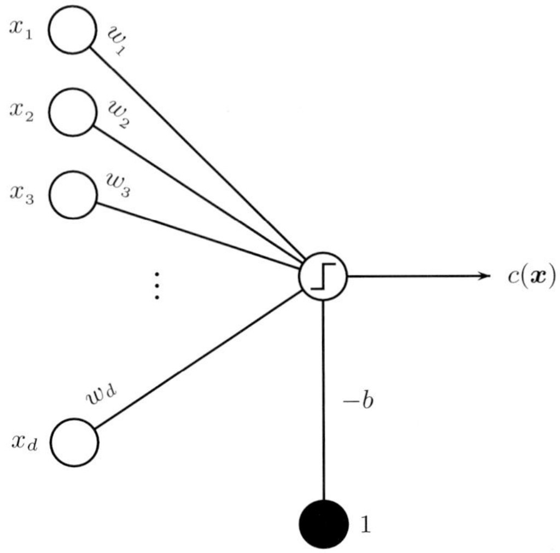

Fig. 4.8. Perceptron with a threshold activation function

4.3.1 Hyperplane Discrimination

This section is concerned mainly with a description of the Perceptron model. However, since its equation is that of a hyperplane, it is important to summarize first the classical Linear Discrimination method developed by [48], as their comparison illustrates why classification methods intended to the search of the class boundaries are more valuable than those based on statistical characterization of the classes.

The classical Fisher's method for discrimination was the first attempt made in Statistics to solve the problem of pattern classification using a hyperplane:

$$g(\boldsymbol{x}) = \boldsymbol{w}^{\mathrm{T}}\boldsymbol{x} - b\,. \tag{4.10}$$

Here $\boldsymbol{w}$ is a vector of weights and b is a threshold. This method can labeled as non-parametric because it does not assume a probabilistic structure of the data. However, it is optimal if the two classes are Gaussian [136]. Its classification rule is as follows: Assign an incoming vector $\boldsymbol{x}$ to $\mathcal{C}_1$ if

$$(\boldsymbol{\mu}_1 - \boldsymbol{\mu}_2)^{\mathrm{T}}\boldsymbol{C}^{-1}\boldsymbol{x} > \frac{1}{2}(\boldsymbol{\mu}_1 - \boldsymbol{\mu}_2)^{\mathrm{T}}\boldsymbol{C}^{-1}(\boldsymbol{\mu}_1 + \boldsymbol{\mu}_2) \tag{4.11}$$

or to $\mathcal{C}_2$ otherwise. Here $\boldsymbol{\mu}_i, i = 1, 2$ is the sample mean vector of class i and $\boldsymbol{C}$ is the pooled covariance matrix computed over the two training sets. Fisher's hyperplane can be put in the form

$$g(\boldsymbol{x}) = \langle \boldsymbol{w}, \boldsymbol{x} \rangle - b = 0 , \tag{4.12}$$

where the weight vector and the bias are given by

$$\begin{aligned} \boldsymbol{w} &= \boldsymbol{C}^{-1}(\boldsymbol{\mu}_1 - \boldsymbol{\mu}_2) , \\ b &= \langle \boldsymbol{w}, \boldsymbol{\mu} \rangle , \end{aligned} \tag{4.13}$$

where $\boldsymbol{\mu}$ is the overall mean. Compare with the FORM solution (1.22). The method is also used for dimensionality reduction. Notice that (4.13) means that the weight vector corresponds to the maximum difference between the class means, normalized by the pooled covariance matrix.

Rosenblatt's Perceptron Algorithm was one of the first milestones in the development of intelligent classification and neural networks [141]. Its equation is also that of a hyperplane. Its fundamentals are as follows: Let us assign a number $c_i \in \{+1, -1\}$ to each sample falling in class $\mathcal{C}_1$ or in class $\mathcal{C}_2$, respectively, and define the *functional margin* $\Delta_i, i = 1, 2, \ldots, n$ for each pattern as

$$\Delta_i = c_i(\langle \boldsymbol{w}, \boldsymbol{x}_i \rangle - b) , \tag{4.14}$$

where $\langle \cdot, \cdot \rangle$ stands for dot product as usual (Fig. 4.7). In words, the functional margin of a sample is defined as the product of the actual class label by the hyperplane equation $\langle \boldsymbol{w}, \boldsymbol{x} \rangle - b$ evaluated at the corresponding sample. Notice that a positive margin indicates right classification of the sample by the hyperplane $(\boldsymbol{w}, b)$, the sign of the class notwithstanding. This formulation contains an important difference with respect to Fisher's method, in that the separating hyperplane is described directly in terms of the samples and not on their statistics.

Rosenblatt [141] developed a simple algorithm for determining $\boldsymbol{w}$. The algorithm attempts to minimize the error function consisting in the sum of misclassifications

$$\begin{aligned} R_{\mathrm{e}}(\boldsymbol{w}) &= \sum_{i \in \mathcal{M}} \Delta_i \\ &= \sum_{i \in \mathcal{M}} c_i(\langle \boldsymbol{w}, \boldsymbol{x}_i \rangle - b) , \end{aligned} \tag{4.15}$$

where $\mathcal{M}$ is the set of misclassified samples by the hyperplane $(\boldsymbol{w}, b)$, i.e. those for which $\Delta_i \leq 0$. The algorithm employs a simple gradient descent technique. Thus the update of the weight vector and the bias b when a sample $\boldsymbol{x}_i$ is presented to the perceptron is

$$\begin{aligned} \boldsymbol{w}[k+1] &\leftarrow \boldsymbol{w}[k] + \eta c_i \boldsymbol{x}_i \,, \\ b[k+1] &\leftarrow b[k] + \eta c_i X^2 \,, \end{aligned} \tag{4.16}$$

where $X^2 = \max_{1 \leq i \leq n} ||\boldsymbol{x}_i||$ and η is a learning parameter whose value is unimportant because the weight vector and the bias can always be re-scaled without affecting the discrimination. It can be shown that the essential convergence properties of Rosenblatt's algorithm depend on the notion of margin (Novikoff's theorem [121]). Hence the margin emerges as a key concept for classification. A condition for using Rosenblatt's algorithm is that the classes must be linearly separable, because otherwise the algorithm never stops. A remedy to this has been proposed under the name of *pocket algorithm* [51]. It yields the best linear approximation for classification of a given set of samples after a finite number of steps.

The applicability of Rosenblatt's Perceptron is limited in the structural reliability practice because its equation is that of a hyperplane. As an illustration, Fig. 4.9 shows the Perceptron decision function for a two-dimensional case in the standard space with the limit state function $g(\boldsymbol{u}) = 2.5 - u_1 - u_2 = 0$. The training set shown in the figure was obtained as follows: One thousand samples were generated after the standard Normal distribution; then the samples having an Euclidean distance to the origin less than 1.5 or lying in other quadrants than the first were removed. This is an application of the general idea for generating training sets for statistical learning devices exposed in Sect. 4.6. No application, however, was made of data compression techniques as suggested there. It can be seen in the figure that the decision line correctly separates the two-class samples. The actual limit-state function is not drawn as it is almost indistinguishable from its estimate. On the contrary, the solution obtained by Fisher's criterion (4.12) is not satisfactory.

In [69] it is argued that, despite its simplicity, the linear Bayesian method known as Linear Discriminant Analysis is one of the best methods for pattern recognition according to some benchmark tests. However, for structural reliability purposes it must be taken into account the criteria for judging the applicability of statistical learning algorithms stated in section 2.12. One of them is the perfect separability of the classes in reliability analysis, which rarely occurs in pattern recognition problems using samples derived from nature. For the case of perfect separability it is intuitively clear that discrimination methods working directly with samples, such as Rosenblatt's Perceptron, are superior over those based on global statistical measures such as Fisher criterion or Bayesian classification. This is because these measures are more decisive when there is a high risk that an incoming sample falls on the wrong class, because some of the training vectors do; however, when this is not possible because of the noise-free physics of the problem (recall that the underlying performance function is deterministic), the boundary between the classes must be the ob-

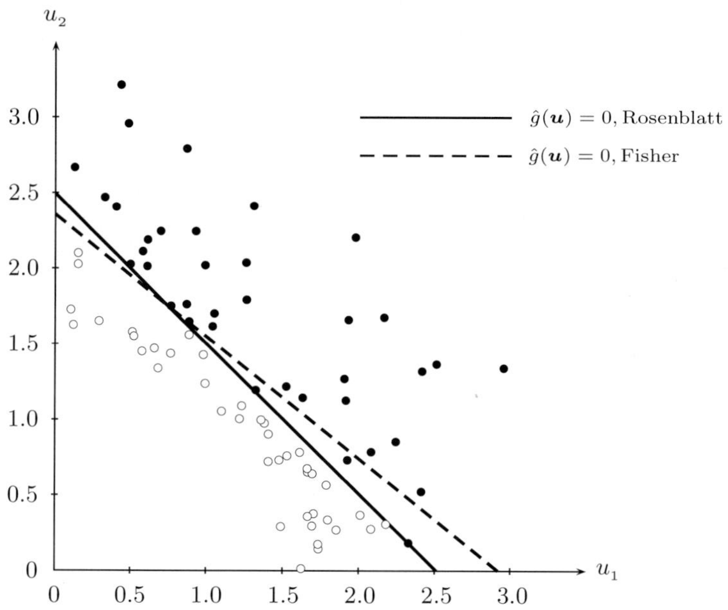

Fig. 4.9. Rosenblatt's Perceptron and Fisher's Criterion solution for a two-dimensional, linear reliability problem

ject of interest. Together with the above discussion on Bayesian classification, the above simple example is another illustration of this fact.

In problems exhibiting high nonlinearity Rosenblatt's Perceptron could eventually be useful for preliminary analyses such as estimating the coordinates of the design point. To this purpose use should be made of the pocket algorithm mentioned before because the samples would not be linearly separable. It is also worth mentioning that a hyperplane separation for structural reliability has been proposed by Ben-Haim in connection with the theory of robust reliability assessment with convex sets [14]. Also, the proposal of [96] for using the RSM is restricted to a hyperplane. In the Support Vector Method examined in the next chapter, such a separation is made possible by generalizing Rosenblatt's notion of the functional margin to the general nonlinear case, but in a different space than that of the basic variables. Hyperplanes in general are considered anew for practical structural reliability analysis in Chap. 7.

4.3.2 Polyhedral Discrimination

The hyperplane discrimination carried out either with Rosenblatt's or Fisher's methods can be represented with the equation

$$c(\boldsymbol{x}) = f\left(\langle \boldsymbol{w}, \boldsymbol{x} \rangle - b\right) , \tag{4.17}$$

where $f(\cdot)$ is a the sign function (also known as *hard limiter* function). This equation is illustrated with the diagram appearing in Fig. 4.8, which is a common way of representing neural networks and related methods. The first layer of neurons is known as the *input layer* while that giving the result is the *output layer*.

Let us now generalize this perceptron in order to build a multi-layer structure that can be more useful for structural reliability purposes. It has been demonstrated [106] that by appending a new layer of neurons (*hidden layer*) with the same activation functions, the resulting structure is capable of discriminating simply connected polyhedra like that shown in Fig. 4.10(**b**). With an additional layer the perceptron can separate two classes whose samples lie in non-connected domains as shown also in the figure. It can be noticed that two-layer structure with hard-limiter functions almost suffices for structural reliability purposes. However, it would be desirable that the perceptron should be able to deal with nonlinear boundaries. This could be achieved by changing the hard-limiter function by a smoother function that operates a nonlinear transformation of its arguments.

Before leaving this section it is interesting to examine whether MLP with hard-limiter activation functions can be used for solving some structural reliability analyses. For instance, polyhedra like that of Fig. 4.10(**b**) appear naturally in problems with linear limit-state functions, like those posed by frames with plastic hinges forming several collapse modes. Also, when hyperplanes are used for several approximations of the reliability of structural systems involving several limit-state functions [40]. However, direct use of 2LP with hard-limiter activation functions does not accommodate well to these problems. In order to illustrate this, consider Fig. 4.11 which displays three limit state functions with a training set. The functions composing the resulting two-layer classifier with hard-limiter functions are shown in Fig. 4.12. It is evident that the number of neurons is high and hence that a sort of pruning is necessary in order to reduce the complexity and the overfitting. After applying a pruning procedure based on the correlation matrix of the estimates the result shown in Fig. 4.13 was obtained. Notice that it has much more functions than those posed by its reliability formulation. Finally, Fig. 4.14 indicates that the training errors and the hyperplanes in excess over the actual

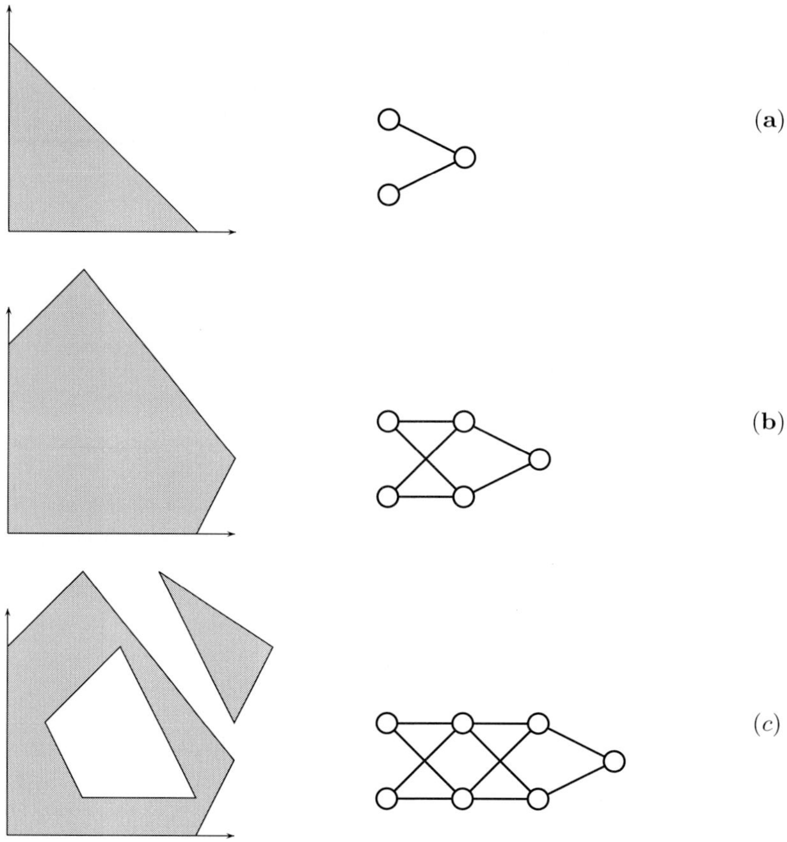

Fig. 4.10. Mappings produced by single- two- and three-layer networks with threshold units. Evidently, two-layer networks suffice for most for reliability analyses treated as classification problems (Inspired in [163])

three functions lead to an erroneous interpretation of the actual class of some failure samples. In particular, there are some intrusions of one class into the other. These intrusions are due to the presence of more lines than actually

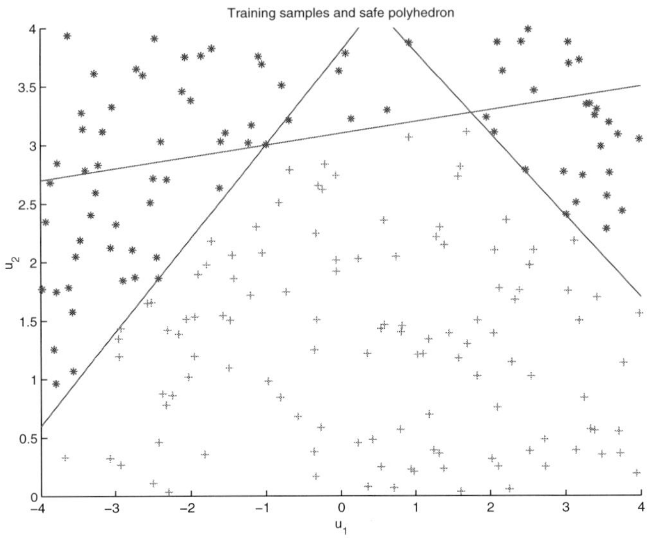

Fig. 4.11. Estimation of multiple reliability problems with 2LP with hard-limiter functions - Training set

needed, which in turn is a consequence of the rigidity of the model. In this respect the method is similar to Classification Trees, though less stiff due to the possibility of fitting tilted planes. However, the example suggests that if there were a means of fitting nonlinear functions with two-layer perceptrons, these problems would be overcome and multiple limit-state functions could be approximated with a single neural network. This can be done with the general nonlinear 2LP.

4.4 General Nonlinear Two-Layer Perceptrons

After the above introduction to neural classification, we are in a position to examine the solution of classification problems with Multi-Layer Perceptrons with other activation functions than the hard-limiter. In the scarce reports on Neural Network applications to solve structural reliability problems, MLP have been the preferred approach and they have been used mainly as a function estimation device [123, 79, 145, 59]. A first proposal to use them as classifiers was made by the author [79] using a perceptron with two layers and sigmoid functions like that shown in Figure 4.15 which according to the above discussion suffices for most structural reliability problems.

In the figure there is an sketch of the transformations usually applied in the neurons of the hidden and output layers. One of them is the logistic sigmoid function, which the most usually employed functions in NN applications. It is given by

$$h(t) = \frac{1}{1 + \exp(-t)} , \tag{4.18}$$

whose domain is $(0, 1)$. Alternatively, use is made of the hyperbolic tangent function

$$h(t) = \frac{\exp(t) - \exp(-t)}{\exp(t) + \exp(-t)} . \tag{4.19}$$

These functions are plotted in Fig. 4.16. The architecture of the network comprises a hidden and an output layers with sigmoid functions. In structural reliability applications more than one neuron could be added in the output layer in order to approximate several limit state functions with a single perceptron. However, the consideration of the reliability problem as a classification task, the preprocessing with techniques such as KPCA (see Chap. 3) as well as the flexibility and adaptivity of the neural networks allow using a single output neuron in most cases. This implies a reduction of the training times and, more importantly, of the training samples.

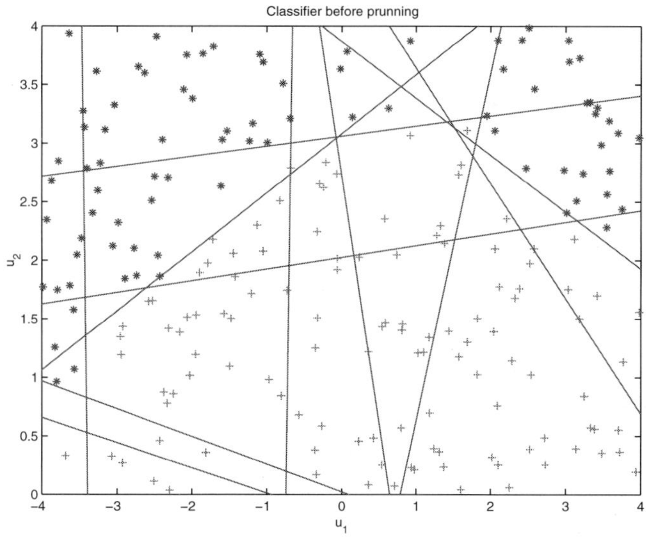

Fig. 4.12. Estimation of multiple reliability problems with 2LP with hard-limiter functions - Classifier before pruning

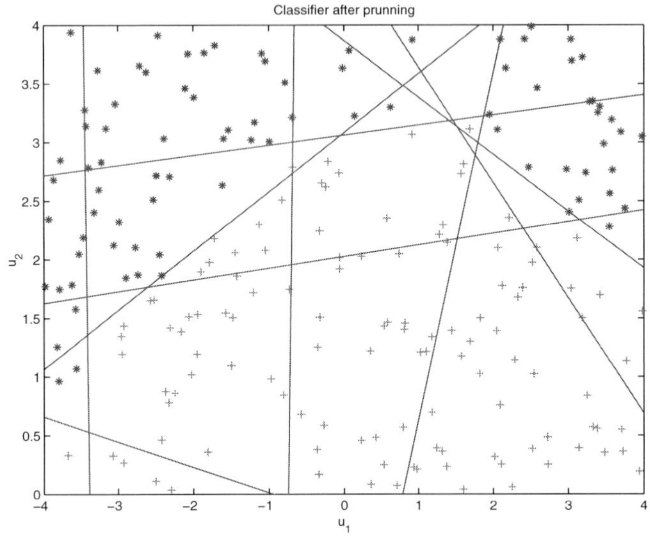

Fig. 4.13. Estimation of multiple reliability problems with 2LP with hard-limiter functions - Classifier after pruning

Let us examine the mapping performed by these networks. If the functions applied in the hidden layer are denoted as $h(\cdot)$ and in the output one as $\tilde{h}(\cdot)$, the estimate of the class labels is

$$c(\boldsymbol{x}) = \tilde{h}\left(\sum_{k=1}^{m} w_k \; h\left(\sum_{i=1}^{d} w_{ki}x_i + w_{k0}\right) + w_0\right) . \tag{4.20}$$

After absorbing the bias terms a more compact form is obtained:

$$c(\boldsymbol{x}) = \tilde{h}\left(\sum_{k=0}^{m} w_k \; h\left(\sum_{i=0}^{d} w_{ki}x_i\right)\right) . \tag{4.21}$$

Finally, in regression applications function $\tilde{h}(\cdot)$ is commonly a purely linear transformation $\tilde{h}(t) = t$. In that case the perceptron is useful for estimating the performance function, i.e.

$$g(\boldsymbol{x}) = \sum_{k=0}^{m} w_{jk} \; h\left(\sum_{i=0}^{d} w_{ki}x_i\right) . \tag{4.22}$$

This equation is identical to that used in linear regression which differs from the conventional one in the kind of the basis functions. In classification applications, in contrast, the use of equation (4.21) with a sigmoid or hard-limiter function at the output units is usually applied in order to obtain a binary result. This implies that the calculation of the weight values is a non-linear optimization programme.

Notice that (4.21) and (4.22) have in common a nonlinear transformation $h(\cdot)$ in the hidden layer and hence an implicit dimensionality increase takes place. Therefore, when used for classification, the equations correspond respectively a nonlinear and a linear discriminating function for the outputs of the hidden layer.

4.4.1 Training Algorithms

Several algorithms have been developed for network training, which are collectively known as backpropagation techniques ([101]; for at-length explanation see [28]). From the numerical viewpoint, they can be categorized as nonlinear optimization strategies, since the target is to find the value of the weight vector $\boldsymbol{w}$ that minimizes a cost function. In most applications use is made of the sum of square errors (SSE) between the actual and estimated outputs. In classification and regression applications the weight updating can be performed pattern by pattern by minimizing the empirical risk function consisting in the sum of squared errors

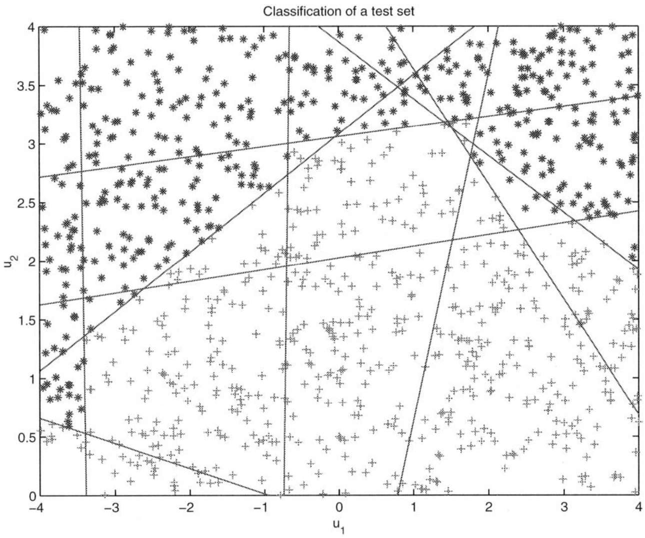

Fig. 4.14. Estimation of multiple reliability problems with 2LP with hard-limiter functions - Test set

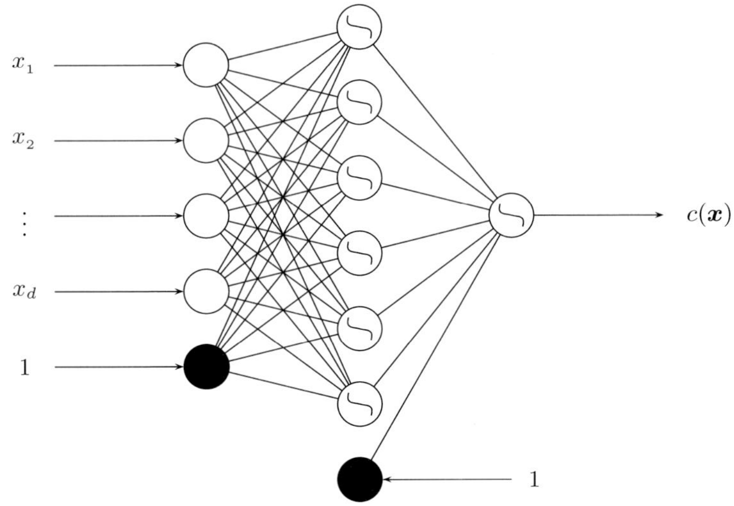

Fig. 4.15. Two-Layer Perceptron useful for reliability analysis treated as a classification task

$$R_{\rm sse} = \frac{1}{2}\sum_{j=1}^{N}\Big(g_j(\boldsymbol{x}) - \hat{g}_j(\boldsymbol{x})\Big)^2 . \tag{4.23}$$

In this equation $g_j(p)$ is the actual output of neuron j at the end layer, $\hat{g}_j(p)$ its MLP estimate, N the number of neurons in the output layer and P the number of training patterns. The output y_j is either a real number, if a sigmoid or a hyperbolic tangent function are used at the output layer, or a binary result if use is made of the hard-limiter. Alternatively, in classification applications only use can be made of the cross entropy function [166]:

$$R_{\rm ent} = \sum_{j=1}^{N} p_j(\boldsymbol{x}) \ln \hat{p}_j(\boldsymbol{x}) + (1 - p_j(\boldsymbol{x})) \ln(1 - \hat{p}_j(\boldsymbol{x})) . \tag{4.24}$$

Here $p_j(\boldsymbol{x})$ denotes the probability that the sample $\boldsymbol{x}$ belongs to either the safe or the failure class, taking the values 1 or 0, respectively, and $\hat{p}_j(\boldsymbol{x})$ is its current estimate. Once the training has been completed by minimizing the average of $R_{\rm ent}(\cdot)$ over all patterns, a sample $\boldsymbol{x}$ is assigned to the failure region if $\hat{p}_j(\boldsymbol{x}) > 0.5$ and to the safe one otherwise.

The weight updating can also be made after all patterns have been shown to the networks and all the estimates have been computed by means of the accumulated errors over all n training patterns:

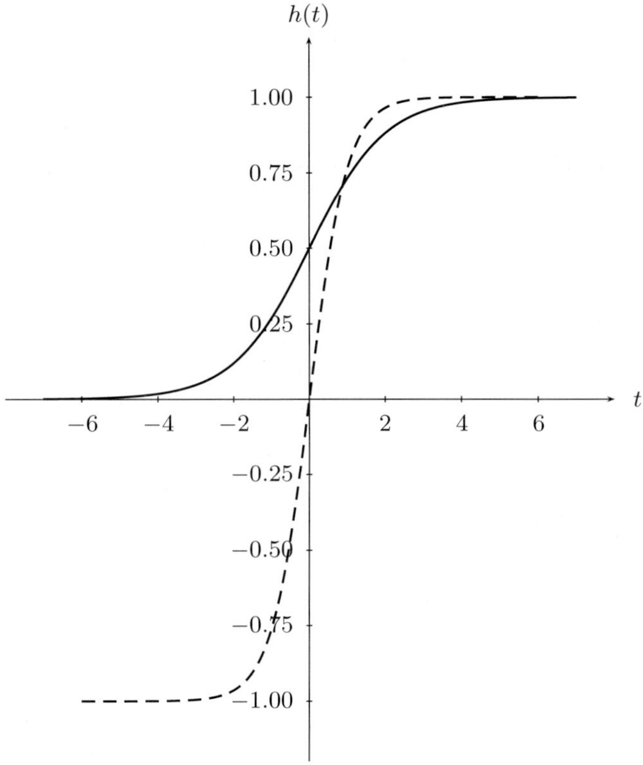

Fig. 4.16. Functions used in Multi-Layer Perceptrons. Solid line: sigmoid. Dashed line: hyperbolic tangent

$$R_{\mathrm{sse}} = \frac{1}{2}\sum_{p=1}^{n}\sum_{j=1}^{N}\Big(g_j(p) - \hat{g}_j(p)\Big)^2 , \tag{4.25}$$

$$R_{\mathrm{ent}} = \frac{1}{2}\sum_{p=1}^{P}\sum_{j=1}^{N} p_j(\boldsymbol{x}) \ln \hat{p}_j(\boldsymbol{x}) + (1 - p_j(\boldsymbol{x})) \ln(1 - \hat{p}_j(\boldsymbol{x})) . \tag{4.26}$$

This later procedure is known as batch-training. The following are the essentials of the most widely applied variants of the back-propagation theme:

1. *Delta rule with variable learning rate and momentum.* This method has its roots in the classical gradient descent technique applied to the single

perceptron (see 4.15. It is described in most books on neural nets. The vector of the network weights is updated at step $[k+1]$ according to

$$\boldsymbol{w}[k+1] = \boldsymbol{w}[k] + \Delta\boldsymbol{w}[k] \;, \tag{4.27}$$

where

$$\Delta\boldsymbol{w}[k] = -\eta\nabla_{\boldsymbol{w}}R[k] + \omega\Delta\boldsymbol{w}[k-1] \;. \tag{4.28}$$

In these equations η and ω are parameters known as *learning rate* and *momentum*, respectively, and $\nabla_{\boldsymbol{w}}R$ is the gradient vector of the empirical risk function R with respect to the weights. The parameters η and ω are set in advance by the user, but some sophisticated commercial codes include routines for their estimation. Other have also the option for making them adaptive, i.e. to modify them as training advances.

2. *Newton-Raphson-Levenberg-Marquardt.* In this case, the weight vector changes with the rule

$$\boldsymbol{w}[k+1] = \boldsymbol{w}[k] - \eta(\boldsymbol{H} + \lambda\boldsymbol{I})^{-1}\nabla_{\boldsymbol{w}}R \;, \tag{4.29}$$

where $\boldsymbol{H}$ is the Hessian matrix of the MLP outputs with respect to the weights.

3. *Gauss-Newton-Levenberg-Marquardt.* In this alternative the vector of weights is updated according to the following equation:

$$\boldsymbol{w}[k+1] = \boldsymbol{w}[k] - (\boldsymbol{J}^{\mathrm{T}}\boldsymbol{J} + \lambda\boldsymbol{I})^{-1}\nabla_{\boldsymbol{w}}R \;. \tag{4.30}$$

Here $\boldsymbol{J}$ stands for the Jacobian matrix of the network outputs with respect to the weights, while $\boldsymbol{I}$ denotes the identity matrix and λ is a convenient nonnegative value.

A theorem by [75] is often quoted as the basis of the MLP methodology for functional approximation. It states that MLP with as few as one hidden layer and arbitrary so-called squashing activation functions and linear or polynomial integration functions can approximate virtually any function provided many hidden units are included in the perceptron. Most activation functions used in MLP uses fall under the class of squashing functions.

4.4.2 Example

Figure 4.17 shows the MLP classifier for the parabolic problem of (4.5) obtained with the 14 training points shown in Figure 4.6 and with the cross-entropy error function (4.24). Logistic sigmoid functions were used in both the hidden and output layers. A very similar result was obtained with the sum of square error function (4.23). It can be seen in the figure that the classifier closely follows the trace of the target limit state function. The steep nonlinear jump illustrates the advantage of using sigmoid functions, as they allow

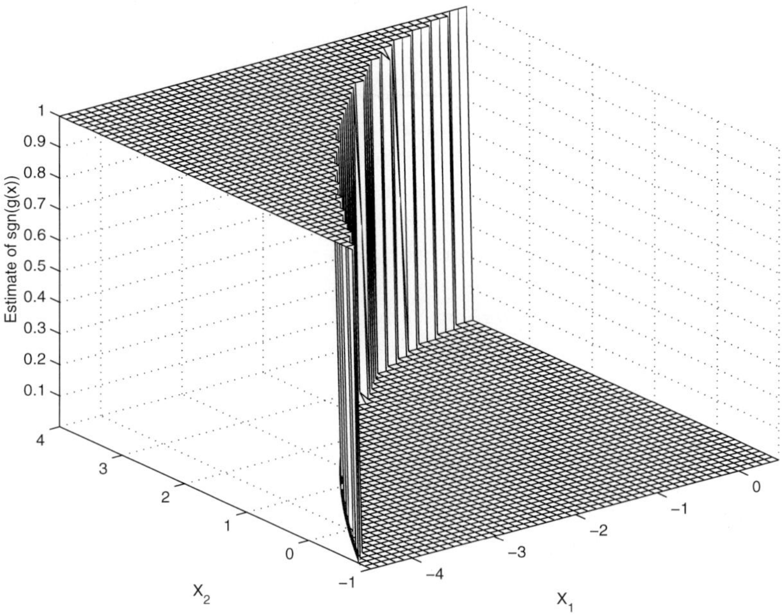

Fig. 4.17. Classifier obtained with a Multi-Layer Perceptron trained with cross-entropy function for a parabolic problem with 14 training points

approximating nonlinear contours while maintaining the hardness of the step functions. The estimates of the probability of failure using both classifiers with 100,000 test samples are equal to 9.4×10^{-4}, which is very close to the exact value, 9.5×10^{-4}. This indicates that the MLP approach is much superior than the Bayesian method and the Classification Tree, as representatives of the probabilistic and Euclidean pattern recognition methods, according to the classification made in Sect. 4.1.

4.4.3 Complexity and Dimensionality Issues

As mentioned in the previous chapter, a quantitative measure of the complexity, the learning capacity and the effects of dimensionality for neural networks is a complex task, since it is difficult to evaluate the VC-dimension of general networks. An important result given in terms of the VC dimension, however, gives a clue on the complexity of neural networks used for classification [4]: For a network with a single hidden layer with sigmoid functions and a single hard-limiter unit as output (a network useful in classification), the VC dimension is bounded by

$$v \leq (WK)^2 + 11WK \log_2(18WK^2) \; , \tag{4.31}$$

where W is the number of free parameters of the network and K the number of computation units. Equation (4.31) means that the VC-dimension of a 2LP grows roughly with the square of the number of connections and processing units, so that the complexity of network's architecture should be subject to control in order to avoid poor generalization properties. Compare this result with the bound for two-layer perceptrons with hard-limiter activation functions in all neurons, that implement polyhedrical discrimination functions, as shown in Fig. 4.10 [4]:

$$v \leq 2W \log_2 \left(\frac{2K}{\ln 2} \right) \; . \tag{4.32}$$

It can be seen that the bound increases roughly one order of magnitude when shifting from hard-limiter to sigmoid functions. This is due to the higher complexity of the sigmoid. For other kinds of networks, especially those applied in regression analysis, important results of this type have been derived in terms of a generalization of the VC-dimension known as *fat-shattering dimension* [4], which consists in allowing for a margin $\gamma \geq 0$ in shattering the classes in such a way that $\gamma = 0$ corresponds to v.

Another important question concerns the effects of the curse of dimensionality in multi-layer perceptrons. The following are some well-established theoretical facts on this issue:

- For a fixed number of training samples, the rate of decrease of the approximation error of neural networks does not depend on d [12]. Particularly, the training error is

$$\text{err} \sim O(1/m) \; , \tag{4.33}$$

  where m is the number of hidden neurons, as in equations (4.20) to (4.22).
- The number of free parameters, however, is bounded by below as [47]

$$W \geq \left(\frac{C_r}{\epsilon} \right)^{d/r} \; , \tag{4.34}$$

  where r is the number of absolutely continuous derivatives of the target function, ϵ the approximation error and C_r a coefficient that depends only on r[1]. The number of nodes is hence, at least,

[1]From the mathematical viewpoint, this supposes that the target function belongs to a Sobolev space.

$$K \geq \frac{1}{d}\left(\frac{C_r}{\epsilon}\right)^{d/r} . \tag{4.35}$$

Equations (4.34) and (4.35) mean that the architecture of the networks depends on the dimensionality of the problem. Notice, however, that if the number of derivatives of the target function is large, $W \to O(1/d)$ independently of the number of dimensions. A result derived by [117] is, perhaps, more interesting from the structural reliability point of view, because it establishes that the number of sigmoidal units is

$$K \sim (d+1)\epsilon^{-d} \tag{4.36}$$

for an approximation within $O(\epsilon)$ of a continuous function. Since the performance function underlying the classification problem belongs to this category, it can hence be expected that the approximation error decreases with the dimensionality in reliability applications.

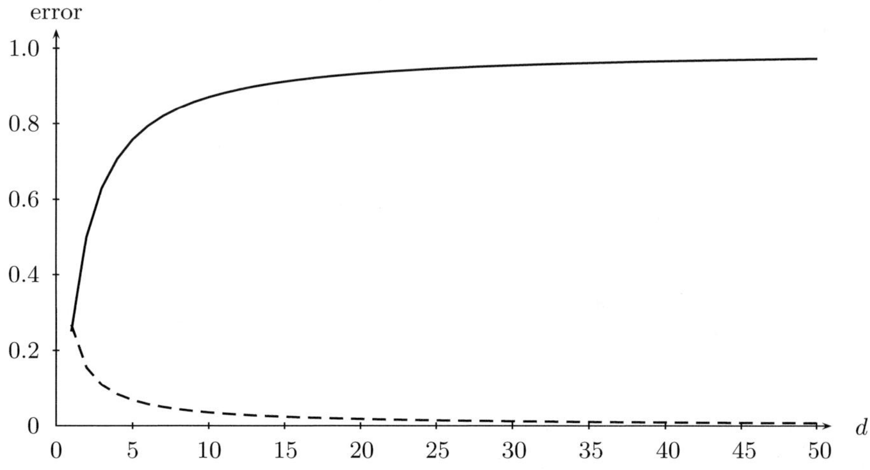

Fig. 4.18. Error in polynomial and neural approximations as a function of dimensionality. The solid line corresponds to the approximation error for a polynomial of degree two, commonly used in the RSM, while the dashed line represents an estimate of the error for a MLP with a number of training samples and hidden layer neurons dependent on the dimensionality

In order to derive some practical conclusions for structural reliability applications of Multi-Layer Perceptrons, it is necessary to delve into these and related equations. On the basis of VC-theory and (4.34), it has been recommended that the number of training samples should be [16]

$$n > \frac{W}{\epsilon} \tag{4.37}$$

for classification purposes. The number of input neurons required in (4.33) is roughly $W/2$ for a single output neuron, which is sufficient for structural reliability analyses regarded as classification tasks. This means that for $\epsilon = 0.1$ about $O(10W)$ training samples are necessary. Now, since the number of hidden layer neurons is normally higher than d, which is the number of input neurons, hence the number of free parameters W increases with the dimensionality. Let us insert these results into a practical recommendation for the number of hidden neurons m [144], which is adapted here for the case of a multi-layer perceptron with one bias in the input layer, one hidden layer and a single output neuron:

$$m = \frac{2 + d + \sqrt{n}}{2} \,. \tag{4.38}$$

For $\epsilon = 0.1$, let $n = 20d$ according to the above reasoning. Therefore, (4.33) yields

$$\text{err} \sim \frac{2}{2 + d(\sqrt{20} + 1)} \,. \tag{4.39}$$

This equation is plotted in Fig. 4.18 together with the training error in a polynomial regression, which is [163]

$$\text{err} \sim O\left(\frac{1}{p^{2/d}}\right) \,, \tag{4.40}$$

where p is the polynomial degree. The plot in the figure corresponds to $p = 2$, which is the order typically used in the Response Surface Method. It is evident that despite the architecture and the number of training samples of Multi-Layer Perceptrons are somewhat dependent on the dimensionality, the end error decreases with this parameter, which is in the spirit of (4.36). In contrast, the error of polynomial classification increases. The empirical evidence of this contrast in the early years of neural network applications stimulated their rapid diffusion and the research in the theoretical demonstration of their learning abilities, some of whose results have been collected in this section.

In practice, it can be said that the MLP can overcome the curse of dimensionality if the architecture of the perceptron and the training set are adequately constituted. In Chap. 5 it is shown that Support Vector Machines provide a gateway to escape from the curse of dimensionality, which is based on their formulation of the optimization problem of the model parameters in a different space from the input one.

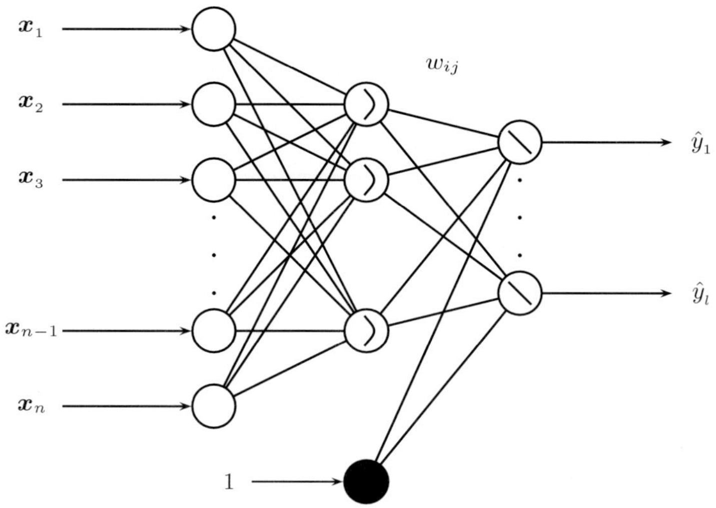

Fig. 4.19. Radial Basis Function network

4.5 Radial Basis Function Networks

Radial Basis Function networks (RBFN) are based on a different concept than MLP. Instead of attempting to establish a global mapping of input onto output variables, i.e. a functional relationship in which to every input corresponds an output like in the MLP case, RBFN networks are intended to perform local mappings, so that the inputs activate the basis functions if they are localized in their active support (so-called *receptive field*); otherwise, their contribution to their activation is close to zero.

Several proposals for the architecture and training algorithms of this network type as well as for the radial basis functions have been published. A typical design is depicted in Figure 4.19. It comprises a first layer, through which all the training vectors $\boldsymbol{x}_l, l = 1, 2, \ldots, n$ are entered. Notice that this makes a difference with respect to MLP, in which the information given to the input neurons is the coordinates of each training pattern. The inputs are directly transmitted to the hidden layer without being affected by weights. The activation in this layer is performed by the radial basis functions described above. Accordingly, in each neuron i, these functions produce the corresponding output according to the proximity of the data to a center $\boldsymbol{\xi}_i$ in the form

$$h_i(\boldsymbol{x}) = h(||\boldsymbol{x} - \boldsymbol{\xi}_i||) . \tag{4.41}$$

A widely used function is the Gaussian one expressed as

$$h(t) = \exp(-\frac{t^2}{2\omega^2}) . \tag{4.42}$$

The constant ω defines the width of the receptive field and is usually assigned beforehand. The output of neuron j of the end layer is computed in a similar way as in the MLP network:

$$\hat{y}_j = f(\sum_{i=1}^{m} w_{ij} h_i(\boldsymbol{x})) , \tag{4.43}$$

where w_{ij} is the weight of the $i \to j$ connection and $f(\cdot)$ is the activation function of the layer. This transformation is usually a purely linear one $f(t) = t$, meaning that

$$\hat{y}_j = \sum_{i=1}^{m} w_{ij} h_i(\boldsymbol{x}) . \tag{4.44}$$

In this case the following problem results for the unknown weights:

$$\boldsymbol{S}\boldsymbol{w} = \boldsymbol{y} . \tag{4.45}$$

Here $\boldsymbol{S}$ denotes the matrix of the radial basis outputs given by (4.41), $\boldsymbol{w}$ is the vector of weights, which are the target of the calculation, and $\boldsymbol{y}$ are the actual outputs. This problem can be formulated basically in two ways, namely, as an exact algebraic problem or as a least squares one. In the first formulation, the size of matrix $\boldsymbol{S}$ is $n \times n$, where n is the number of training patterns. This implies that there must be also n receptive field centers, which are usually the same training points arranged randomly. Therefore the radial basis functions will be activated according to the relative Euclidean distance of the data points to each other. This method normally implies a large number of neurons in the hidden layer. In the second approach, the size of the matrix is $n \times m$, where $m << n$ is the number of hidden neurons (see Equations 4.43 and 4.44). A strategy usually applied is to sequentially increase the size of the hidden layer by adding a single neuron in each step and solving the following least squares problem:

$$\boldsymbol{w} = (\boldsymbol{S}^{\mathrm{T}}\boldsymbol{S})^{-1}\boldsymbol{S}^{\mathrm{T}}\boldsymbol{y} . \tag{4.46}$$

The procedure stops when the square error downcrosses an error measure such as the well-known Akaike Information Criterion. In practice, it is important to balance the relative advantages of these two approaches: the first requires the solution of a large system of equations but it is exact at least for the training set; the second proceeds greedily to reach a minimal network that can be very fast in the production phase, but at the expense of some loss of accuracy with respect to the training set. From the point of view of learning theory, the first approach normally leads to a model of higher complexity than the second, thus putting into risk the generalization capacity. For this reason the second is normally preferred.

4.5.1 Approximation Properties

Similarly to the MLP case, the approximation of functions via radial basis atoms has been theoretically justified and grounded. In particular, it has been demonstrated that they are also universal approximators [66] and that they have a property of best approximation not shared by MLP [57]. This is hardly surprising, as RBFN has the structure of a linear model with basis functions of localized active support like several other models of the kind that show good approximation properties in one dimension, like wavelets for instance.

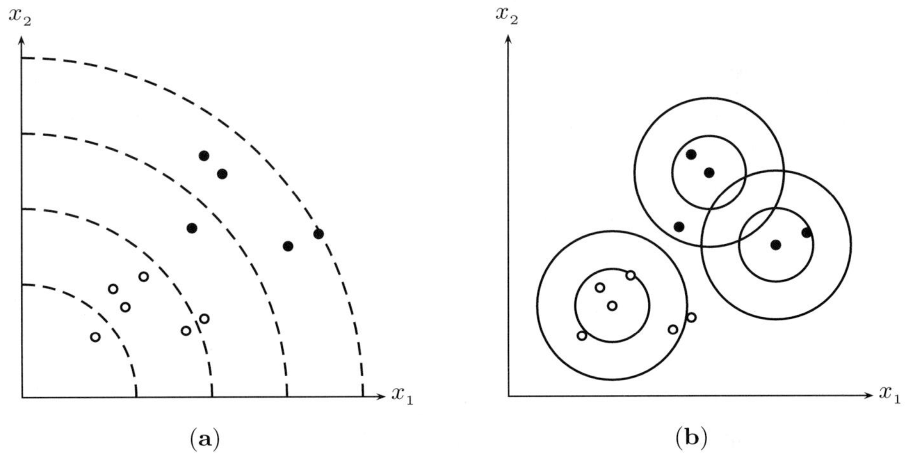

Fig. 4.20. Global and local mappings performed by Multi-Layer Perceptrons (**a**) and Radial Basis Function Networks (**b**)

4.5.2 A First Comparison of MLP and RBFN

There are significant differences between the mappings performed by MLP and RBFN, which contribute to explain why the former are more useful for solving structural reliability problems under the classification viewpoint.

Multi-Layer Perceptrons perform a nonlinear global mapping. In fact, the input to a neuron in the hidden layer is the linear combination $\sum w_j x_j, j = 1, 2, \ldots, d$ so that the output of the neuron is the same for all vectors $\boldsymbol{x}$ such that $\sum w_j x_j = a, a \in \mathbb{R}$, i.e., for all vectors lying in a hyperplane. On the opposite, the RBFN perform several local mappings in which the importance of each data point is given by its proximity to the center of a receptive field, in the manner illustrated by Figure 4.20. In particular, the output $h_i(\boldsymbol{x}) = h(||\boldsymbol{x} - \boldsymbol{\xi}_i||)$ (see 4.42) is the same for all vectors $\boldsymbol{x}$ lying at the same Euclidean distance from centers $\boldsymbol{\xi}_i$. Therefore, a training point is relevant to the end value of some connections stemming from a center if this is close to it. This supports their use in unsupervised-supervised training mode for regression purposes, as described in Chap. 6.

Now, the type of mapping performed by the MLP make them more convenient for classification approach because the contours appearing in Fig. 4.20 simply become parallel ones as in Fig. 4.17 by using a logistic sigmoid or similar function in the output layer neurons. On the contrary, the local mappings performed by radial basis functions are convenient to model density functions like those shown in Figs. 4.3 and 4.4, which are employed in the Bayesian method of classification. Since this method has been discarded for structural reliability applications, it is concluded that RBFN are not convenient in this form. The same is valid for a variant of RBFN known as Probabilistic Neural Network [172], whose only difference with RBFN is that the maximization of the posterior probabilities is performed by a so-called competitive neuron.

Rigorously, the inadequacy of RBFN for the classification needed in structural reliability can be demonstrated by setting the Bayesian classification rule in the form of a linear model like that applied by the RBFN (see 4.44). To this end, the density functions $p(\boldsymbol{x})$ and $p(\boldsymbol{x}|\mathcal{C}_k)$ in (4.1) are formulated as the mixtures [16]

$$p(\boldsymbol{x}) = \sum_k p(\boldsymbol{x}|\mathcal{C}_k)P(\mathcal{C}_k) \,,$$

$$p(\boldsymbol{x}|\mathcal{C}_k) = \sum_{i=1}^{m} p(\boldsymbol{x}|i)P(i|\mathcal{C}_k) \,. \tag{4.47}$$

With these arrangements (4.1) becomes

$$P(\mathcal{C}_k|\boldsymbol{x}) = \sum_{i=1}^{m} w_{ik} h_i(\boldsymbol{x}) \,, \tag{4.48}$$

where

$$w_{ik} = P(\mathcal{C}_k|i) = \frac{P(i|\mathcal{C}_k)P(\mathcal{C}_k)}{P(i)}$$

$$h_i(\boldsymbol{x}) = P(i|\boldsymbol{x}) = \frac{p(\boldsymbol{x}|i)P(i)}{\sum_{\ell=1}^{m} p(\boldsymbol{x}|\ell)P(\ell)} \,. \tag{4.49}$$

This confirms that the RBFN constitutes an adaptive-flexible method for estimating the posterior probabilities required by the Bayesian classification approach and hence they are of little use for structural reliability under the pattern recognition paradigm. On the contrary, the universal approximation property of RBFN make them convenient for being applied in a regression setting for estimating either specific failure probabilities or the distribution function of structural responses. This is not only due to these properties but also because they can be trained in a mixed unsupervised-supervised mode that can eventually render computational savings according to [16].

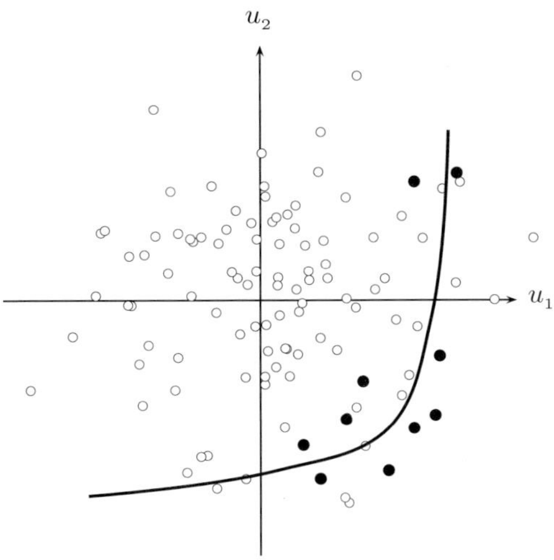

Fig. 4.21. Sampling for training learning devices. The white circles indicate unlabeled samples forming a pool generated with the given distributions. The black circles are labeled samples obtained after selecting samples from the data bank and compressing the resulting set

4.6 Elements of a General Training Algorithm

At a difference to the Response Surface Method, the training of the classification and regression devices which are the subject of this book is made with random samples. In this section, some ideas on a general algorithm to that purpose, which were briefly introduced in Sect. 2.12 are developed.

In practical reliability problems there is no information on the design point or other guidelines for starting the training procedure. The only information is normally the orthant in the standard variable space in which the important part of the limit-state function lies, which in many cases can be inferred upon elementary mechanical considerations. For instance, in the classical load/capacity problem, the performance function crosses the standard two-dimensional space $\boldsymbol{u}$ in the fourth orthant, as illustrated by Fig. 4.21. This will be called the critical orthants. Anyhow, the failure samples, which in this algorithm are the first to be obtained as explained next, constitute in themselves an indication on the orthants of this class.

On this reduced information basis, the statistical learning devices, used either in the classification or regression modes, should be trained using a reduced data set of samples lying on both sides of the limit-state function. Since this is unknown at this stage of the analysis, it is necessary to generate the reduced set on heuristic basis as follows. As is well known, structural failure probabilities are normally very low. This important feature can be used for generating a training population about the limit state function. In fact, given a rough estimate of the failure probability $\hat{P}_{\mathrm{f}} = O(10^{-k})$, one can say that M out of N samples would approximately lie in the failure region, so that $2M$ samples can be said to lie in the neighborhood of the limit state function. Accordingly, the following algorithm can be used for simulating this $2M$ population:

- Map the d basic variables onto the standard space of uncorrelated Normal variables using a suitable transformation (see Section 1.3).
- Simulate a population of the set of basic variables with the multidimensional Normal density function. The number N_1 of random numbers in this phase of the algorithm should be very large because the farthest samples will be used to detect the orthants where failure samples lie. (The orthants can be easily identified with the vector of signs of the random variates). Therefore, a large N_1 renders unlikely a failure occurring in an orthant where the farthest sample yields a safe state.
- Sort the samples according to their distance from the origin:

$$a_j = \sqrt{\sum_{i=1}^{d} u_{ij}^2} \; . \tag{4.50}$$

- Select the last k samples and evaluate the corresponding responses using the structural solver, where k is a small number. Mark the orthants where the performance function is negative.
- Select the last $M = 2N_2\hat{P}_{\mathrm{f}}$ samples lying in the critical orthants as the population surrounding the limit state function, where N_2 should be taken much less than N_1 in order to reduce the number of additional calls of the structural solver.
- Calculate the responses corresponding to the M samples using the solver.

Alternatively, the samples can be selected on the basis of an estimated range of the reliability index $\beta \in [\beta_1 \;\; \beta_2]$. Anyhow, a pool of unlabeled samples

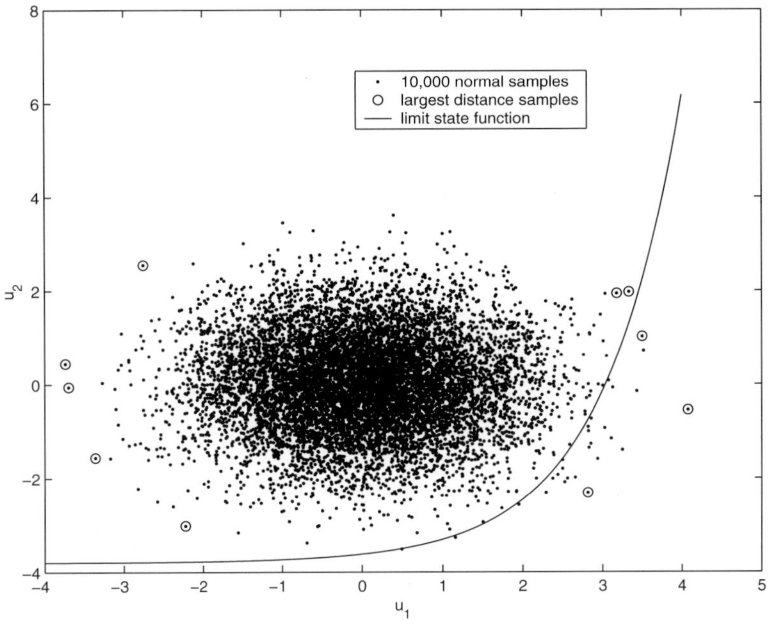

Fig. 4.22. Example of sampling for training learning algorithms. Initial scan with the farthest samples

(i.e. samples for which the performance function or its sign are unknown, because they have not been processed by any solver) is obtained.

As an illustration, consider Fig. 4.22 that displays the limit state function

$$g(u_1, u_2) = -3.8 + \exp(u_1 - 1.7) - u_2 = 0 \tag{4.51}$$

and $N_1 = 10,000$ standard Normal samples. Those lying the farthest from the origin are framed with circles. They indicate that the critical orthants are the first and fourth. Then a polulation of $N_2 = 1,000$ samples was generated and the twelve farthest samples lying in the critical orthants are selected for training a surrogate of the limit state function (Fig. 4.23).

Inasmuch as labeling each sample has a high computational cost in some cases, it is desirable to reduce this set further for training the devices in cases where each calculation of the performance function takes much computation time. Such small subset of the pool would then be labeled by the solver. In order to make this selection some strategies could be applied:

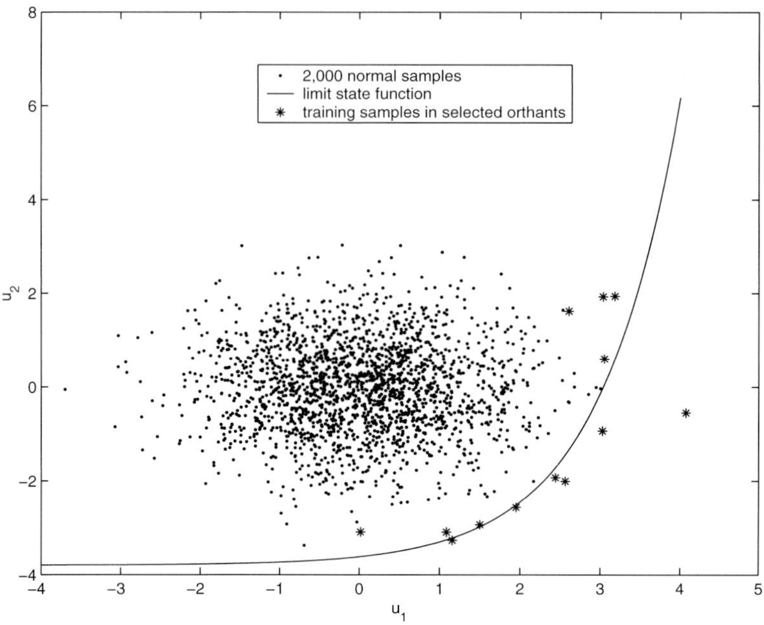

Fig. 4.23. Example of sampling for training learning algorithms. Training samples in selected orthants

- *Random selection.* This has the inconvenient that the presence of training samples on both domains (safe and failure), which is absolutely necessary for starting the training algorithm, is not guaranteed. Matching this condition by random selection is especially difficult in high-dimensional spaces.
- *Compression techniques.* The final training samples are obtained by data compression techniques such as Vector Quantization or the centers of a kernel density model calculated with the Expectation-Maximization algorithm. With these techniques the above drawback of random selection can be easily overcome because they perform an optimal data compression as shown in Chap. 3. This is therefore the recommended procedure.

With the exception of the last step, this algorithm was used in generating the training set for the simple perceptron decision function shown in Fig. 4.9.

Before closing this section, it must be said that other heuristic techniques could be applied, such as sequential greedy selection of the training samples out of the pool. In Support Vector Machines such a selection is made possible by the availability of explicit equation for a margin, beyond which no selection of new training samples is necessary if it correctly embraces the limit-state function. On this basis some algorithms for SVM training was developed. They are exposed in the next chapter.

5

Classification Methods II - Support Vector Machines

5.1 Introduction

This chapter is devoted exclusively to the classification approach to structural reliability using Support Vector Machines, which at a difference to Neural Networks were developed apart from biological paradigms and were instead construed entirely in the framework of Statistical Learning Theory. For structural reliability applications they show significant advantages over other classification methods for at least five reasons:

- They include a powerful instrument, called the *margin*, that can be exploited to reduce the number of finite element calls.
- The apply the SRM (Structured Risk Minimization) inductive principle, described in Chap. 2, that for perfectly separable problems grants a null empirical risk together with the best solution amongst those allowed by the number of training samples.
- At a difference to neural networks, they do not require nonlinear optimization techniques for finding the model weights. Instead, the maximization of the margin leads to a simple Quadratic Programming problem for whose solution standard techniques can be applied.
- Very importantly, they can overcome the curse of dimensionality in a more explicit manner than Multi-Layer Perceptrons. Such overcoming is made possible both in the search of model parameters and in the VC-dimension that controls complexity and generalization errors. The former occurs by a change of the search domain from the primal space of basic variables to the dual space of certain Lagrange multipliers. The latter, by the properties of the optimal hyperplane in the feature space. As a consequence, the number of training samples and the generalization capacity of the model can be made independent of the number of dimensions.
- Their mathematical formulation with a hyperplane in high dimensions can be used as a basis for the formulation of a robust reliability index for cases with scarce probabilistic information (see Chap. 7).

The chapter is organized as follows. First, a detailed description of the fundamentals and basic features of Support Vector Machines is presented. Special regard is paid to the problems of complexity and dimensionality. Then genetic and active learning algorithms for solving reliability problems with SVM are proposed. Their application is illustrated with some examples.

Historically, the Support Vector Machine was the first statistical learning algorithm that proposed the use of Mercer kernels for classification, regression and density estimation. Its success in these tasks promoted the development of other kernel-based algorithms such as the KPCA and Kernel Fisher Discriminant for feature extraction and classification methods like Relevance Vector Machines, Bayes Point Machines and others. In this monograph, the Support Vector Machine has been selected among them for structural reliability applications because it is a well established method for which there is abundant experience in several fields. However, for the sake of completeness of the exposition other kernel-based classification algorithms are summarized at the end of the chapter.

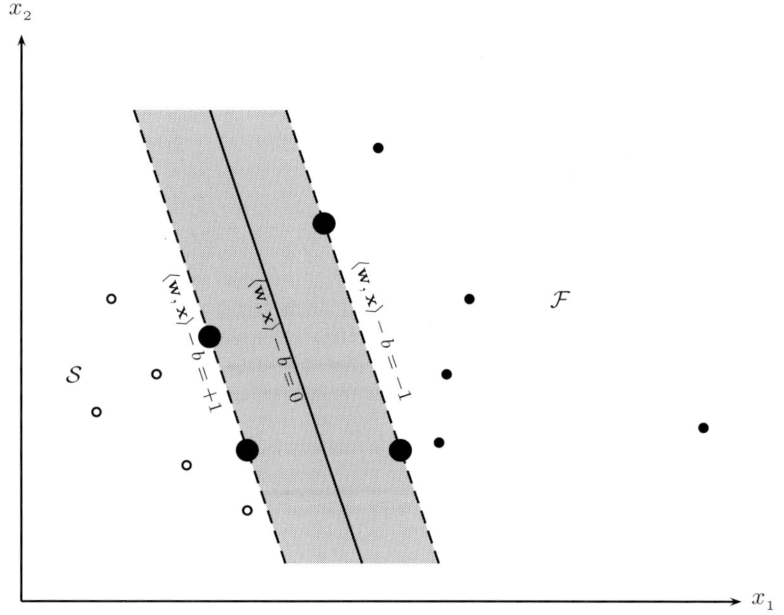

Fig. 5.1. Linear separation of classes by optimal hyperplane. The support vectors are the big black circles

5.2 Support Vector Machines

This section is devoted to a summary of the support vector method for pattern recognition in the specific case of separable classes, which is more relevant to structural reliability than the non-separable task. However, this latter constitutes the basis of the support vector approach for regression and therefore it is summarized in the next chapter.

5.2.1 Linearly Separable Classes

Let us consider first the problem of linearly separable classes (see Figure 5.1). Let the patterns be $\{\boldsymbol{x}_1, \boldsymbol{x}_2, \ldots, \boldsymbol{x}_n\}$ and the classes to which they belong be identified as $c_i \in \{-1, 1\}, i = 1, 2, \ldots, n$. A separating hyperplane of the form

$$g(\boldsymbol{x}) = \langle \boldsymbol{w}, \boldsymbol{x} \rangle - b = 0 \tag{5.1}$$

is sought, where $\boldsymbol{w}$ is a vector of parameters that define the normal vector to the hyperplane, and b is threshold. The condition imposed to this hyperplane is that it maximizes the distance to the given patterns, as this allows the best confidence on the classification it performs. Thus, the optimization problem reads

$$\max_{\boldsymbol{w}, b} \min_i \{||\boldsymbol{x} - \boldsymbol{x}_i|| : \langle \boldsymbol{w}, \boldsymbol{x} \rangle - b = 0, i = 1, 2, \ldots, n\} \, . \tag{5.2}$$

The definition of the classes through a sign facilitates the formulation of this optimization problem. In fact, the constraints for the preceding optimization problems are

$$\begin{aligned} \langle \boldsymbol{w}, \boldsymbol{x}_i \rangle - b &\geq +1, \text{ if } c_i = +1 \, , \\ \langle \boldsymbol{w}, \boldsymbol{x}_i \rangle - b &\leq -1, \text{ if } c_i = -1 \, , \end{aligned} \tag{5.3}$$

which can be compactly put in the form of Rosenblatt's functional margin (4.14)

$$\Delta_i = c_i(\langle \boldsymbol{w}, \boldsymbol{x}_i \rangle - b) \geq 1 \, . \tag{5.4}$$

It is recalled that a positive result indicates right classification, the sign of the class notwithstanding. By normalizing the parameters of the hyperplane by $1/||\boldsymbol{w}||$, it is concluded that the points on the margin satisfy $|\langle \boldsymbol{w}, \boldsymbol{x} \rangle - b| = 1$, as shown in Figure 5.1. The normalization is needed because there are infinitely many solutions for the vector $\boldsymbol{w}$ of the optimal hyperplane that differ only in a scale factor. Let us define the *hard margin* as the distance from the hyperplane to the closest point in a separable case (A *non separable* case is defined by the existence of points for which $c_i(\langle \boldsymbol{w}, \boldsymbol{x}_i \rangle - b) < 1$ and it makes use of a so-called *soft margin*, as explained in Chap. 6). After normalization the maximum margin (i.e, the maximum distance from the hyperplane to the closest sample) becomes

$$\Delta = \frac{1}{||\boldsymbol{w}||} , \tag{5.5}$$

as can be easily demonstrated. In fact, substracting the equation for one of the side-hyperplanes $\langle \boldsymbol{w}, \boldsymbol{x}^- \rangle - b = -1$ from that of the other $\langle \boldsymbol{w}, \boldsymbol{x}^+ \rangle - b = +1$ yields $\langle \boldsymbol{w}, (\boldsymbol{x}^+ - \boldsymbol{x}^-) \rangle = 2$, which upon dividing by the norm of the vector $\boldsymbol{w}$ and averaging yields (5.5).

The optimization problem in the primal space of hyperplane parameters $(\boldsymbol{w}, b)$ can be stated as

$$\text{minimize } \Omega(\boldsymbol{w}) = \frac{||\boldsymbol{w}||^2}{2}$$

$$\text{subject to: } c_i(\langle \boldsymbol{w}, \boldsymbol{x}_i \rangle - b) \geq 1, i = 1, 2, \ldots, n.$$

This means that minimizing the length of the weight vector under constraints (5.4) is equivalent to maximizing the margin, thus allowing a good capacity for right classification of incoming samples not used in the training phase. The square of the length is used as the function to be minimized in order to cast the problem in a standard quadratic programming form. The optimization problem can be recast as

$$L(\boldsymbol{w}, b, \boldsymbol{\alpha}) = \frac{1}{2} \|\boldsymbol{w}\|^2 - \sum_{i=1}^{n} \alpha_i [c_i(\langle \boldsymbol{w}, \boldsymbol{x}_i \rangle - b) - 1] \tag{5.6}$$

via Lagrange multipliers $\alpha_i \geq 0$. Hence, the problem is to find a saddlepoint because it is necessary to minimize the loss function with respect to the hyperplane parameters $(\boldsymbol{w}, b)$ while maximizing it with respect to the Lagrange multipliers. This implies solving

$$\frac{\partial L(\boldsymbol{w}, b, \boldsymbol{\alpha})}{\partial b} = 0 ,$$

$$\frac{\partial L(\boldsymbol{w}, b, \boldsymbol{\alpha})}{\partial \boldsymbol{w}} = 0 . \tag{5.7}$$

The solution is

$$\sum_{i=1}^{n} \alpha_i c_i = 0$$

$$\boldsymbol{w} = \sum_{i=1}^{n} \alpha_i c_i \boldsymbol{x}_i . \tag{5.8}$$

Taking into account the positiveness of the Lagrange multipliers set forth above, (5.8) have a very important meaning for the structural reliability problem: The separating hyperplane can be expressed only in terms of the patterns that have a positive Lagrange multipliers, while the rest are no longer needed[1]. This is due to the fact that if the problem constraint holds strictly, i.e. $c_i(\langle \boldsymbol{w}, \boldsymbol{x} \rangle - b) > 1$, then the value of α_i that maximizes the cost function in (5.6) is necessarily $\alpha_i = 0$. Such is the so-called Karush-Kuhn-Tucker complementarity condition of optimization theory which reads (e.g. [89]):

$$\alpha_i [c_i(\langle \boldsymbol{w}, \boldsymbol{x}_i \rangle - b) - 1] = 0, i = 1, 2, \ldots, n \, . \tag{5.9}$$

The special patterns having $\alpha_i > 0$ are the support vectors, which will be denoted as $\boldsymbol{x}_j, j = 1, 2, \ldots, s$. They have the important property of lying just on the margins defined above, as the preceding equation clearly shows. In words, this means that they are the points most difficult to classify. Its defining property is crucial for the development of the sequential Monte Carlo algorithm reported in [80] and its further developments exposed in the sequel, since it implies that the only samples needed to update the classifier are those lying in the band defined by the margins.

The above exposition indicates that the solution for the weights of the hyperplane can be expressed as a convex combination of some input patterns in the form

$$\boldsymbol{w} = \sum_{j=0}^{s} \alpha_j c_j \boldsymbol{x}_j \, , \tag{5.10}$$

where s is the number of support vectors. An interpretation of this equation is that the optimal hyperplane can defined only in terms of some patterns that are close to it, which agrees with intuition.

For passing into the nonlinear separation, it is convenient to cast the optimization problem in terms of the so-called dual variables, i.e. the Lagrange multipliers. This is done by replacing (5.8) into (5.6):

maximize

$$W(\boldsymbol{\alpha}) = \sum_{i=1}^{n} \alpha_i - \frac{1}{2} \sum_{i=1}^{n} \sum_{j=1}^{n} \alpha_i c_i \langle \boldsymbol{x}_i, \boldsymbol{x}_j \rangle c_j \alpha_j \tag{5.11}$$

subject to:

---

[1] It has been observed [148] that equations 5.8 are analogue to those of basic mechanics: The first corresponds to the equilibrium of forces of magnitude α_i and direction c_i applied to a rod. The second to the equilibrium of moments, because $\sum \boldsymbol{x}_i \times (\alpha_i c_i \boldsymbol{w}) = \boldsymbol{w} \times \boldsymbol{w} = 0$.

$$\alpha_i \geq 0 \,,$$
$$\sum_{i=1}^{n} \alpha_i c_i = 0 \,. \tag{5.12}$$

The solution of this quadratic optimization problem should be substituted into (5.10) to obtain the values of the hyperplane parameters. The threshold b can then be obtained from the Karush-Kuhn-Tucker condition as exposed in Sect. 5.2.2. The classification function then reads

$$c(\boldsymbol{x}) = \text{sgn}(g(\boldsymbol{x})) = \text{sgn}\left(\sum_{i=1}^{n} \alpha_i c_i \Big(\langle \boldsymbol{x}_i, \boldsymbol{x} \rangle - b\Big)\right) \tag{5.13}$$

where $s < n$.

Before closing this section it is important to call the reader's attention to the fact that the optimization problems in primal and dual forms (5.6 and 5.11, respectively) differ in that the solution to the former should be sought in $\boldsymbol{w} \in \mathbb{R}^d$, where d is the dimensionality of vector $\boldsymbol{x}$, while the latter in the domain of the vector of Lagrange multipliers $\boldsymbol{\alpha} \in \mathbb{R}^s$, where s is the number of support vectors[2]. This feature constitutes a remarkable difference of SVM over the approximations of the limit-state function calculated with FORM, SORM, RSM and Neural Networks. Therefore, if $s < d$ the use of the dual formulation makes the problem independent of the dimensionality, which in some structural reliability problems can be very high. As shown next, this inequality is easily enforced by a kernel nonlinear projection that produces a dimensionality explosion, which renders highly positive effects for hyperplane classification (see Section 2.10). This is a first reason explaining why SVM can overcome the curse of dimensionality. The second one is a consequence of the feasibility of hyperplane separation in a large number of dimensions exposed in Sect. 2.10. Such a dimensionality increase is induced by this method in the way described next.

5.2.2 Nonlinear Separation

Nonlinear separation using support vector machines is based on the positive effects of dimensionality increase for hyperplane discrimination exposed in Section 2.10. Such an increase is achieved by projecting the samples onto a higher dimensional space (called *feature space*), while maintaining a hyperplane formulation. For instance, if $\boldsymbol{x}$ is a vector in a space of dimensionality d with coordinates $(x_1, x_2, \ldots, x_d)$, the projection on a feature space with a polynomial mapping of degree two yields a vector with $D = d(d+3)/2$ coordinates [175]:

[2]Rigorously, $\boldsymbol{\alpha} \in \mathbb{R}^n$, where n is the number of training samples. However, since most elements of the vector $\boldsymbol{\alpha}$ are null the intrinsic dimensionality of the vector is s.

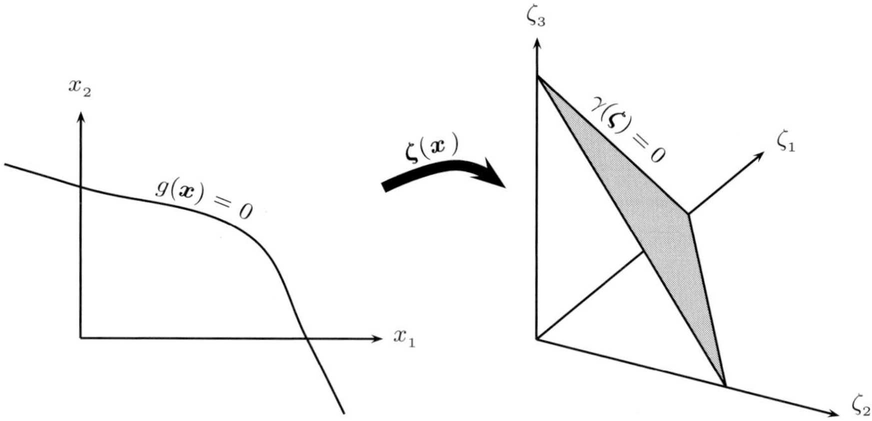

Fig. 5.2. Nonlinear mapping for support vector classification

$$\begin{aligned} d \text{ coordinates} &: \zeta_1 = x_1, \ldots, \zeta_d = x_d \,, \\ d \text{ coordinates} &: \zeta_{d+1} = x_1^2, \ldots, \zeta_{2d} = x_d^2 \,, \\ \frac{d(d-1)}{2} \text{ coordinates} &: \zeta_{2d+1} = x_1 x_2, \ldots, \zeta_D = x_{d-1} x_d \,. \end{aligned} \tag{5.14}$$

Hence, with the mapping $\boldsymbol{\zeta} : \mathbb{R}^d \mapsto \mathbb{R}^D$ (5.13) takes the form

$$c(\boldsymbol{x}) = \operatorname{sgn}(g(\boldsymbol{x})) = \operatorname{sgn}\left(\sum_{i=1}^{n} \alpha_i c_i \Big(\langle \boldsymbol{\zeta}(\boldsymbol{x}_i), \boldsymbol{\zeta}(\boldsymbol{x}) \rangle \Big) - b \right) . \tag{5.15}$$

The nonlinear projection (see figure 5.2) is easily done by generalizing equation (5.15) to

$$c(\boldsymbol{x}) = \operatorname{sgn}(g(\boldsymbol{x})) = \operatorname{sgn}\left(\sum_{i=1}^{n} \alpha_i c_i K(\boldsymbol{x}_i, \boldsymbol{x}) - b \right) , \tag{5.16}$$

where the linear kernel in (5.13) has been substituted by the general kernel

$$K(\boldsymbol{x}, \boldsymbol{y}) = \langle \boldsymbol{\zeta}(\boldsymbol{x}), \boldsymbol{\zeta}(\boldsymbol{y}) \rangle \,, \tag{5.17}$$

in which the vector function $\boldsymbol{\zeta}(\cdot)$ represents the nonlinear mapping. The key point in this kernel presentation is that the nonlinear mapping $\boldsymbol{\zeta}(\cdot)$ need not be known explicitly, as indicated previously in the discussion of KPCA (see

Sect. 3.3). As it was observed there, if the kernel satisfies the conditions posed by Mercer's theorem it can have virtually infinite dimensions, thus greatly facilitating the hyperplane separation in the feature space, according to the exposition on the benefits of dimensionality increase exposed in section 2.10. This makes the kernel approach a powerful tool for solver substitution in the reliability context because the implicit large dimensionality implies a very high probability of correct classification using a hyperplane, according to the above said.

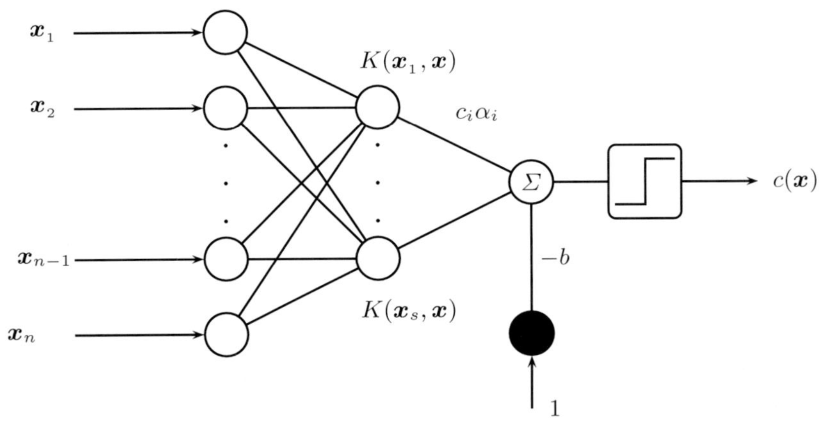

Fig. 5.3. Network representation of a Support Vector Machine

In terms of the kernel, the optimization problem becomes

maximize

$$W(\boldsymbol{\alpha}) = \sum_{i=1}^{n} \alpha_i - \frac{1}{2}\sum_{i=1}^{n}\sum_{j=1}^{n} \alpha_i c_i K(\boldsymbol{x}_i, \boldsymbol{x}_j) c_j \alpha_j \tag{5.18}$$

subject to:

$$\alpha_i \geq 0 \,,$$

$$\sum_{i=1}^{n} \alpha_i c_i = 0 \,. \tag{5.19}$$

Also, note that in the feature space the equation for the weights is

$$\boldsymbol{w} = \sum_{i=1}^{n} \alpha_i c_i \boldsymbol{\zeta}(\boldsymbol{x}_i) , \tag{5.20}$$

but this calculation, that requires the explicit knowledge of the nonlinear mapping, is not actually needed because the classification function is given by the dual form (5.16). The threshold can be obtained from the Karush-Kuhn-Tucker condition (5.9), formulated directly in terms of the kernel as

$$b = \sum_{i=1}^{n} \alpha_i c_i K(\boldsymbol{x}_i, \boldsymbol{x}_j) - c_j, \;\; j = 1, 2, \ldots, s . \tag{5.21}$$

In practice, the s values of the threshold resulting from the application of this equation are averaged. However, the values are quite similar if the margin is very narrow, as is the case in structural reliability applications due to the infinitesimal transition from the safe to the failure class.

Figure 5.3 displays a network representation of the Support Vector Machine. It has been represented such that the hidden layer contains much less processing units than the input layer, which is in contrast to multi-layer perceptrons.

Finally, for some ensuing developments, it is important to remark that in the feature space the equation of the hyperplane is

$$g(\boldsymbol{\zeta}) = \langle \boldsymbol{w}, \boldsymbol{\zeta} \rangle - b . \tag{5.22}$$

5.2.3 Solution of the Optimization Problem

In principle, the optimization problem given by (5.18) and (5.19) can be solved with standard QP (Quadratic Programming) techniques. These, however, are avoided in pattern recognition application because in this field the number of training samples is normally very large. For this reason, some special algorithms have been developed [126, 93, 148]. In structural reliability, on the contrary, the purpose is to train the machines with a minimal number of samples. For this reason, the applicability of QP routines is straightforward.

To this end let us put the SVM problem in the general QP form:

$$\begin{aligned} &\text{minimize } \frac{1}{2}\boldsymbol{\alpha}^{\mathrm{T}} \boldsymbol{H} \boldsymbol{\alpha} + \boldsymbol{q}^{\mathrm{T}} \boldsymbol{\alpha} \\ &\text{subject to } \mathbf{0}_n \leq \boldsymbol{\alpha} \mathbf{1}_n , \end{aligned} \tag{5.23}$$

where $\boldsymbol{\alpha}$ is the vector of Lagrange multipliers, $\mathbf{0}_n, \mathbf{1}_n$ are vectors of length equal to the number of samples n having zero and one as all its elements, respectively. In this formulation

$$\boldsymbol{H}_{ij} = c_i c_j K(\boldsymbol{x}_i \boldsymbol{x}_j) \,, \tag{5.24}$$

$$\boldsymbol{q} = -\boldsymbol{1}_n \,. \tag{5.25}$$

Notice that the calculation of the parameters of a SVM is much easier than those of a MLP, especially if this latter has nonlinear functions at the output layer. The main difference between the two approaches in this respect lies in the fact that the convex problem of SVM has one global solution, while that of MLP with nonlinear functions in the output layer can have many local minima on which the required nonlinear optimization procedures can become trapped.

5.3 A Remark on Polynomial Chaoses

As has been shown, Support Vector Machines implement the idea that hyperplane separation is highly likely in a large dimensional space by projecting the samples to it using a kernel function. It is interesting to notice that a similar goal can be accomplished by means of the polynomial chaos expansion summarized in Sect. 1.3.5.

Our interest here is to observe that this expansion allows the decomposition of any random variable into a large number of Gaussian variates. Therefore, if in a reliability problem of d dimensions this expansion is performed for each random variable $x_j, j = 1, 2, \ldots, d$, the number number of dimensions becomes much larger, thus facilitating the separation of the safe and failure domains through a hyperplane, according to Cover's theorem. To be specific, if the expansion for each variable is applied in the form

$$\underline{x}_j = a_0^{(j)} \Gamma_0^{(j)} + \sum_{i_1=1}^{p} a_{i_1}^{(j)} \Gamma_1(\underline{u}_{i_1}^{(j)}) + \sum_{i_1=1}^{p} \sum_{i_2=1}^{i_1} a_{i_1 i_2}^{(j)} \Gamma_2(\underline{u}_{i_1}^{(j)}, \underline{u}_{i_2}^{(j)}) + \ldots \,, \tag{5.26}$$

then one has that the new space of random variables has $d' = d \times p$ dimensions.

It can be noticed that despite this approach could render a similar benefit as that obtained with Support Vector Machines, the projection onto a space of a very large number of dimensions is attained in a more simple manner using the kernel trick applied by SVM. Notice also that the dimensionality increase does not occur if the expansion of the random variable is made in terms of one-dimensional Hermite polynomials, which is a method that has been applied in structural reliability [183, 40].

5.4 Genetic Algorithm

As stated above, the purpose of using a SVM in reliability analysis is to derive a criterion for separating the samples lying in the safe and failure domains.

To this end it is necessary to find a separating hyperplane in the feature space whose mapping onto the input space approaches the limit-state function as close as possible. An algorithm devised to such a purpose was reported in [80]. It belongs to the family of so-called *active learning* algorithms explained in the next section. In this section a genetic approach is proposed. Despite it is also an instance of an active learning, it is exposed apart from this general family because of the special procedures of biological inspiration that it employs.

5.4.1 General Considerations

The following are some general considerations for a sequential support vector approximation of the limit-state function. A SVM might be trained over a large set of samples obtained by any Monte Carlo sampling procedure that parsimoniously scans the entire input space, such as Latin Hypercube Sampling [49]. However, the very concept of support vectors makes unnecessary this anyhow costly approach, as no other vectors than the support ones enter into the definition of the hyperplane in the feature space, as shown before. Hence, the notion of support vectors can be exploited for minimizing the number of function calls $g(\boldsymbol{x})$, which is commonly the result of a finite element calculation. Assuming that a pool of samples has been created according to the steps of the general algorithm of Sect. 4.6, the updating of the current SV classifier can then be made by generating samples lying in the current margin, because those lying beyond it give no new information, as their Lagrange multiplier will be zero. A sample $\boldsymbol{u}$ is in the margin of a given SVM if

$$\left| \sum_{i=1}^{n} \alpha_i c_i K(\boldsymbol{u}_i, \boldsymbol{u}) - b \right| < 1\,. \tag{5.27}$$

In order to generate samples in this zone, it is proposed to use the underlying pool equipped with a few hundreds of samples of the basic variables, generated from their actual joint density function or any other convenient choice. These samples are not classified, i.e. they are not processed by the numerical solver of the structural model, as they simply constitute an underlying population on which the SV classifier will be sequentially updated. After training the first classifier with a small subset of samples selected with data compression techniques as stated in section 4.6, the result can be used to initiate a genetic algorithm for a sequential improvement follows.

5.4.2 Algorithm

As is well-known, Genetic Algorithms are a gradient-free method for function minimization, which have their inspiration in the laws of the evolution of species. To reach this goal, GA generate sequential populations of samples (called generations) and on each of them apply the following basic operations (see e.g. [58]):

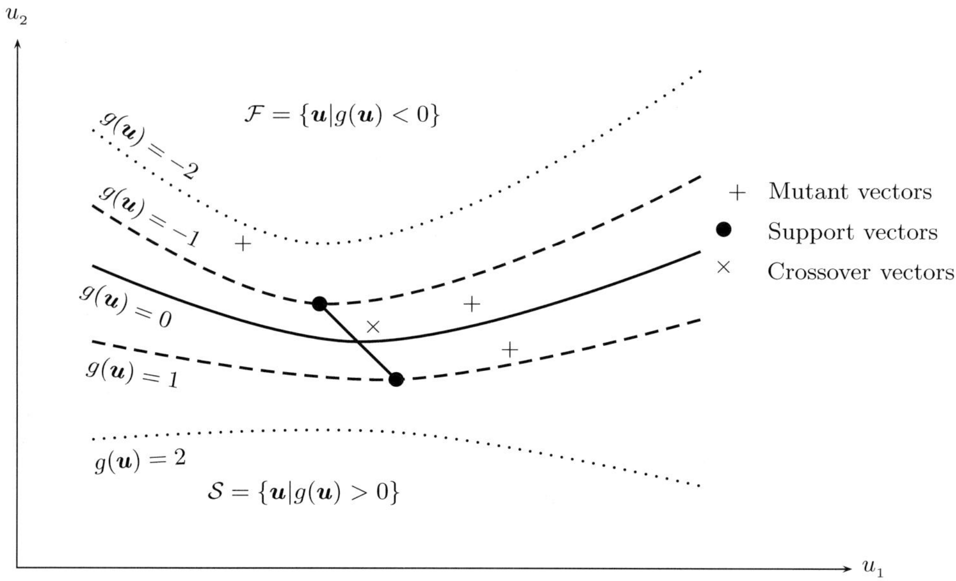

Fig. 5.4. Genetic algorithm for Support Vector Machine updating

- *Selection.* The individuals composing each generation are tested by the cost function and only the best are retained for constituting the seed of a future generation.
- *Crossover.* New individuals of the next generation are produced by crossing the genes of the best individuals from the former.
- *Mutation.* In order to avoid the degeneracy of the species that can be a result of the above two operations, some new individuals are created by mutation of the chromosomes of the best individuals of the preceding population.

In their early developments, the GA operated on binary strings of the variables in order to imitate the structure of chromosomes. The Evolutionary Strategies, which were developed in parallel, were intended to work with real numbers, but the procedures are similar [11]. Nowadays, these two currents have converged to a single field of optimization methods which have in common their biological inspiration and either name is used indistinctly to denote this philosophy. In this book we have adopted the classical name of genetic algorithms.

Although there are many variants of this scheme, the above are the essential procedures applied to each generation. The following GA procedures can be applied for a sequential approximation of the implicit function:

- *Selection.* According to the above description of Support Vector Machines, it is evident that the solution of the extreme problem (5.18) can be regarded as equivalent to the evaluation of the cost function after generating a population in the GA. More importantly, since the support vectors are the only samples that satisfy the Karush-Kuhn-Tucker condition (5.9), they are just the best of their generation in the GA sense. In other words, this condition is in itself a selection mechanism. In addition, as shown in the next section, Support Vector Machines naturally implement the Structured Risk Minimization inductive principle by which it is granted that the best hyperplane (in the sense of minimum generalization error when using the classifier) in the feature space is the one given by the solution of the optimization problem (5.18). Therefore, the selection phase in the current generation is already solved by the calculation of the support vectors.
  It is expected that the support vector representation stemming from the next iteration outperforms the previous ones in that the new margin is smaller, due to the following definitions of the crossover and mutation procedures. For this reason they are used as the parent generation.
- *Crossover.* The production of new individuals on the basis of the support vectors can be done in several ways. For instance, it is tempting to combine two support vectors lying in different sides of the boundary function by searching the zero crossing of the performance function using simple linear interpolation. However, this produces linearly dependent vectors which would create numerical difficulties in solving the QP problem (5.18). For this reason it is better to apply a random method. It is proposed to use linear combinations of the form

$$\begin{aligned} \boldsymbol{u}_j &= \xi \boldsymbol{s}^{(\mathcal{C}_1,*)} + (1-\xi)\boldsymbol{s}_j^{(\mathcal{C}_2)} + \epsilon ||\boldsymbol{s}^{(\mathcal{C}_1,*)}||\boldsymbol{v} \ , \\ \boldsymbol{u}_k &= \xi \boldsymbol{s}^{(\mathcal{C}_2,*)} + (1-\xi)\boldsymbol{s}_k^{(\mathcal{C}_1)} + \epsilon ||\boldsymbol{s}^{(\mathcal{C}_2,*)}||\boldsymbol{v} \ . \end{aligned} \tag{5.28}$$

  Here $\boldsymbol{s}_i^{(\mathcal{C}_i)}$ denotes any support vector in class $\mathcal{C}_i, i = 1, 2$ and the star denotes the best current approximations to the design point over the current set of support vectors in both classes. These approximations are obtained by simply selecting the support vectors that are the closest to the origin in $\mathcal{C}_1$ and $\mathcal{C}_2$. On the other hand, $\xi, \underline{\boldsymbol{v}}$ are a uniform random number and a uniform random vector in the ranges $[0, 1]$ and $[-1, 1]$, respectively, and ϵ is a small fixed number. The addition of a small portion of the norm of the design point approximation with random direction cosines grants that the new samples lie in whereabouts of the line joining the best individuals of the previous generation, but not necessarily on it because in that case the crossover equation would have the form $\boldsymbol{u}_j = \eta \boldsymbol{s}^{(\mathcal{C}_1)} + (1-\eta)\boldsymbol{s}_j^{(\mathcal{C}_2)}$

and the problem of linear dependence would be induced. Therefore, the crossover vectors (5.28) can be regarded as genetic descendants to the support vectors. Let us denote the number of these individuals as c.

- *Mutation.* As said above, in the standard binary GA algorithms, the mutants are generated by distorting some chromosomes of the best individuals. In our case, due to the availability of the functions defining the margin, the mutants can be generated by producing m random samples related to the margin in some way. Since the crossover points are generated inside the margin, the mutants could be used to explore new regions beyond it, in order to prevent that the approximation becomes wrongly closed in a zone of the variable space. Taking into account that the margin has a unit normalized width, the mutation operation is performed by selecting m random samples from a band equal to twice the margin. This is also illustrated by Fig. 5.4. In that way the mutants can be regarded as heirs to the previous support vectors in a weaker sense than are the samples resulting from the crossover operation.

According to this algorithm, the number of individuals composing the next generation is $s+c+m$, from which only $c+m$ function evaluations should be done in each step because the functions corresponding to the support vectors had already been evaluated. Evidently, other evolutionary strategies could be applied to produce a sequential approximation of the implicit function. However, the simple genetic approach described in this section interprets correctly the elitist nature of the support vector methodology.

Once the training is finished, the failure probability P_{f} is assessed using

$$\hat{P}_{\mathrm{f}} = \frac{1}{2} - \frac{1}{2M}\sum_{i=1}^{M} \mathrm{sgn}\left[g(\boldsymbol{u}_i)\right] , \tag{5.29}$$

where M is the number of samples in the production phase, i.e. when the SVM acts as a solver substitute. This equation stems from the fact that the conventional $\{0, 1\}$ classification used in reliability analysis via the indicator function is substituted by the $\{-1, 1\}$ classification performed by the signum function.

5.5 Active Learning Algorithms

Many pattern recognition methods are applied in a passive way. This means that they calculate the classifier on the basis of a given set of labeled data. There are some areas in which it is important to select the training samples from a pool of unlabeled data, because the labeling operation of all the available samples is difficult [42]. This is the case, for instance, in which labeling is costly because it depends upon an expert, a lab experiment, a soil exploration, etc. Also, when the amount of unlabeled samples is so large that it is necessary to devise an algorithm for selecting the training samples.

In structural reliability, training a classifier with random samples can be an instance of the first category of problems, namely when labeling the sample depends on a large numerical solution. However, despite in some cases the labels can obtained very fast, the low value of the failure probability implies that most samples belong to the same class and it is therefore necessary to avoid obtaining redundant information. For these reasons, active learning is a relevant approach in this field.

The main differences of active learning (also known as *learning with queries*) with passive learning are, therefore, that the information is given in the form of a pool of unlabeled data and that there is a selecting function that picks up the next query to improve the current classifier. For structural reliability applications the pool can be generated by the method proposed in Sect. 4.6.

The next paragraphs expose two active learning algorithms. The first was reported in [80]. The second is a general algorithm for active learning proposed by [170]. Section 5.8 is devoted to application examples of the active as well as genetic algorithms.

5.5.1 Algorithm Based on Margin Shrinking

The algorithm reported in [80] is oriented to a sequential shrinking of the margin. It is as follows:

MARGIN SHRINKING ALGORITHM
$t = 0$;
// *Generate a pool of unlabeled patterns*
(i.e. realizations of the input variables $\boldsymbol{x}$*)*
initialize $Q(t) = \{\boldsymbol{x}_1, \boldsymbol{x}_2, \ldots, \boldsymbol{x}_N\}$;
// *Initialize the training set:*
// *Let* $\boldsymbol{x}_\mathrm{s}$ *and* $\boldsymbol{x}_\mathrm{f}$ *be a safe point and a failure point, respectively*
initialize $Q_\mathrm{tr}(t) = \{(\boldsymbol{x}_\mathrm{s}, -1), (\boldsymbol{x}_\mathrm{f}, +1)\}$;
$STOP = 0$;
while $(STOP == 0)$
  $t = t + 1$;
  $f_t(\boldsymbol{x}) =$ **train SVM** $(Q_\mathrm{tr}(t-1))$;
  // *Look for the nearest point to the separating function*
  $i = \underset{\forall i:\, \boldsymbol{x}_i \in Q(t)}{\arg\min} \; (|f_t(\boldsymbol{x}_i)|)$;
  if $(|f_t(\boldsymbol{x}_i)| < 1)$ **then**
    $c_i = 2\Upsilon\left[g(\boldsymbol{x}_i) \le 0\right] - 1$;
    // *Augment the training set and reduce the pool*
    $Q_\mathrm{tr}(t) = Q_\mathrm{tr}(t-1) \cup \{\boldsymbol{x}_i, c_i\}$;
    $Q(t) = Q(t-1) \backslash \{\boldsymbol{x}_i\}$;
  else
    // *The margin band is void*

$STOP = 1$;
end if
end while

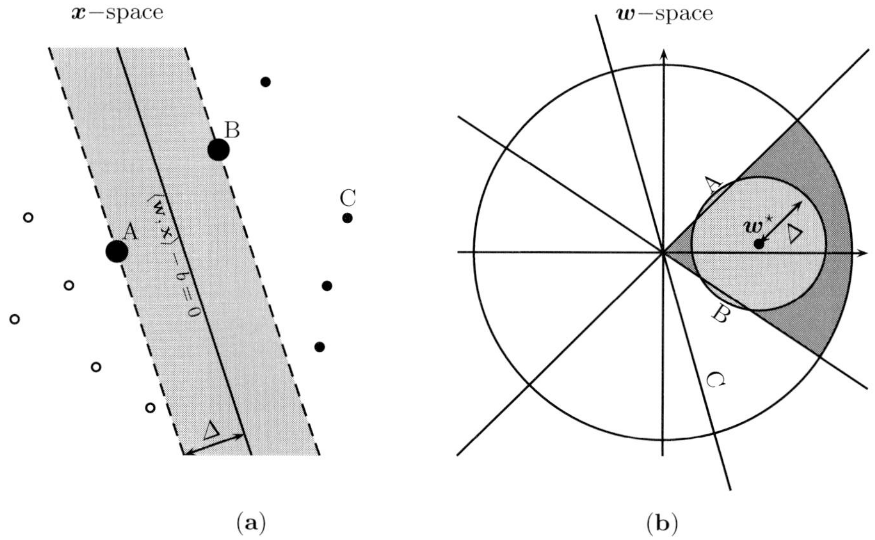

Fig. 5.5. Version space. (**a**) Linear support vector solution in the input parameter space. (**b**) The version space is the dark area in the weight space determined by the hypersphere $||\boldsymbol{w}|| = 1$ and the hyperplanes of normal vectors $c_i \boldsymbol{x}_i$ corresponding to the support patterns. The SVM solution given the current training samples is the inner hypersphere of radius equal to the margin, centered at the optimal solution $\boldsymbol{w}^\star$

5.5.2 Algorithm Based on Version Space Shrinking

This algorithm was developed by [170]. It hinges upon the concept of *version space*, which is explained first. Let us consider the basic support vector problem, i.e. that of the linear discrimination of perfectly separable problems. Equation 5.1 can be regarded either as the equation of a hyperplane of coefficients $\boldsymbol{w}$ in the $\boldsymbol{x}$−space or as that of a hyperplane of coefficients $\boldsymbol{x}$ in the $\boldsymbol{w}$−space. Since in the SVM method the weights $\boldsymbol{w}$ are normalized such that $||\boldsymbol{w}|| = 1$, the search domain of the weights is a hypersphere in the $\boldsymbol{w}$−space (see Fig. 5.5). Consequently, each sample $\boldsymbol{x}_i$ splits the hypersphere into two halves, one that correctly classifies the sample (i.e. $\langle \boldsymbol{w}, \boldsymbol{x}_i \rangle = c_i$) and another that does not. For this reason, the normal to the hyperplane $c_i\boldsymbol{x}_i$ oriented by the sample class points to the halfsphere that contains the solution for the weights. The intersection of all these correct halfspheres, limited by the the planes $\boldsymbol{x}_i$ is called the *version space*. On the other hand, it can be easily shown that the SVM solution can be represented in the $\boldsymbol{w}-$space by another hypersphere, of radius equal to the half margin, that touches the hyperplanes of the support vectors (Fig. 5.5). See [143] for the derivation an other details.

In [170] it is demonstrated that the SVM solution $\boldsymbol{w}^\star$ lies in the hyperplane that halves the surface area occupied by the hyperpsphere $||\boldsymbol{w}|| = 1$ in the version space. Accordingly, they have developed three algorithms for searching this optimal solution using indicators of the ratio between the two parts of such an area. The algorithms operate a shrinkage of the version space about the optimal solution and therefore the convergence to it is guaranteed. One of them is the so-called MaxMin margin algorithm, in which the indicator is the minimum margin of the test patterns computed with the current classifier assuming positive and negative labels. It is adapted herein to the structural reliability problem as follows:

MAXMIN MARGIN ALGORITHM

$t = 0$;
//Generate a pool of unlabeled patterns
(i.e. realizations of the input variables $\boldsymbol{x}$*)*
initialize $Q(t) = \{\boldsymbol{x}_1, \boldsymbol{x}_2, \ldots, \boldsymbol{x}_n\}$;
// Initialize the training set:
// Let $\boldsymbol{x}_{\mathrm{s}}$ *and* $\boldsymbol{x}_{\mathrm{f}}$ *be a safe point and a failure point, respectively*
initialize $Q_{\mathrm{tr}}(t) = \{(\boldsymbol{x}_{\mathrm{s}}, -1), (\boldsymbol{x}_{\mathrm{f}}, +1)\}$;
$STOP = 0$;
while $(STOP == 0)$
  $t = t + 1$;
  $f_t(\boldsymbol{x}) =$ **train SVM** $(Q_{\mathrm{tr}}(t-1))$;
  for $i = 1 : n - t - 2$
  // Test pattern assuming +1 class with the current SVM:
  ${}^t\Delta_{+1}(\boldsymbol{x}_i) = {}^t\,\mathrm{SVMmargin}(Q_{\mathrm{tr}}(t-1), +1)$
  // Test pattern assuming -1 class with the current SVM:
  ${}^t\Delta_{-1}(\boldsymbol{x}_i) = {}^t\,\mathrm{SVMmargin}(Q_{\mathrm{tr}}(t-1), -1)$
  end for
// Select the best candidate for query:
$j = \underset{\forall i:\, \boldsymbol{x}_i \in Q(t)}{\arg\min}\ ({}^t\Delta_{+1}(\boldsymbol{x}_i), {}^t\Delta_{-1}(\boldsymbol{x}_i))$
if $t \leq n - 2$
// Augment the training set and reduce the pool
$Q_{\mathrm{tr}}(t) = Q_{\mathrm{tr}}(t-1) \cup \{\boldsymbol{x}_j, y_j\}$;
$Q(t) = Q(t-1) \backslash \{\boldsymbol{x}_j\}$;
else
  // The margin band is void
  $STOP = 1$;

end if
end while

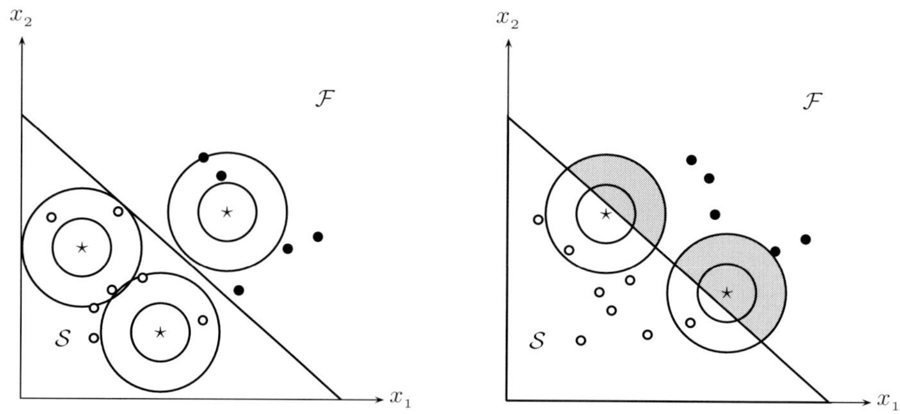

Fig. 5.6. Differences between Radial Basis Function Networks and Support Vector Machines using Gaussian kernels. In RBFN the centers can be made optimal with respect to samples, while in SVM they are optimal with respect to the boundary. In addition, SVM includes the class sign as a factor in the model, which accounts for the class change across the boundary

5.6 A Comparison with Neural Classifiers

In the introducing discussion on Statistical Learning devices made in Chap. 1 it was stated that Multi-Layer Perceptrons are recommended to be used either as classifiers or as regressors for the reliability problem while the use of Radial Basis Function Networks is only recommended in the latter form (see Table 2.1). This is due to the fact that when used for classification they become estimators of the Bayes posterior probability as said in the preceding. The implementation of both kinds of Neural Networks for functional approximation is the subject of the next chapter.

It can be somewhat surprising that two models having a similar structure, namely RBFN and a SVM using a Radial Basis kernel, can be so differently useful for solving the classification problem posed by Structural Reliability. In order to clarify this, consider Figure 5.6, which shows these to models. The RBFN model assigns an input $\boldsymbol{x}$ to the class having

$$\max_k \left(\sum_{i=1}^{m} w_{ik} K(\boldsymbol{x}_i, \boldsymbol{x}) - b \right) , \tag{5.30}$$

while the SVM renders a sign label for the assessed class :

$$\mathrm{sgn} \left(\sum_{i=1}^{n} \alpha_i c_i K(\boldsymbol{x}_i, \boldsymbol{x}) - b \right) . \tag{5.31}$$

The following are the differences between them:

- In RBFN the centers are sometimes made optimal with respect to the available training samples, using techniques such as Vector Quantization or Expectation Maximization (See Chap. 3) for determining the centers; also, regularization or supervised learning approaches can be applied to that purpose. In SVM, in contrast, the optimization is made with respect to the boundary between the classes. For this reason the centers of the kernel functions are points close to the boundary while in RBFN they can be far from it.
- In RBFN the weights are class probabilities and hence they are positive. Also are the basis functions as such. In SVM the Lagrange multipliers, which are constrained to be positive, are multiplied by the class sign. As a consequence, despite the activation of samples lying at a given distance from the Radial Basis Function center is the same regardless of their position, in SVM a sudden jump is enforced to this output as the boundary is crossed by the radial basis kernel. Therefore, the uniformity of radial function activation and its probabilistic interpretation are altered by the introduction of the class sign.
- The RBFN model is guided by a Bayesian approach and hence no use is necessary of the sign function. Instead, the supremum operator is applied to select the highest posterior probability. In SVM the probabilistic approach is abandoned in favor of a geometrical one. For this reason, despite the similarities of the models (5.30) and (5.31), the SVM model plays a similar role to that of MLP, i.e. as a global approximation of the limit state function as a threshold. This explains why they yield better results in Structural Reliability problems than probabilistic methods, when trained with a small set of samples lying in the vicinity of the limit-state function, as already shown.

To close this section it is necessary to discuss the application of the training procedures introduced above to Multi-Layer Perceptrons. At a difference to SVM, the MLP do not have the powerful instrument of the margin that allows a selection of new training samples for refining the approximation in a sequential approach. Therefore it is necessary to apply an heuristic algorithm such as the distance-based approach explained in Chap. 4. The practical issues of Neural Network and Support Vector Machine applications for the solution of the full probabilistic problem (i.e. the estimation of $\hat{P}_{\underline{g}}(g)$) are examined in the next chapter.

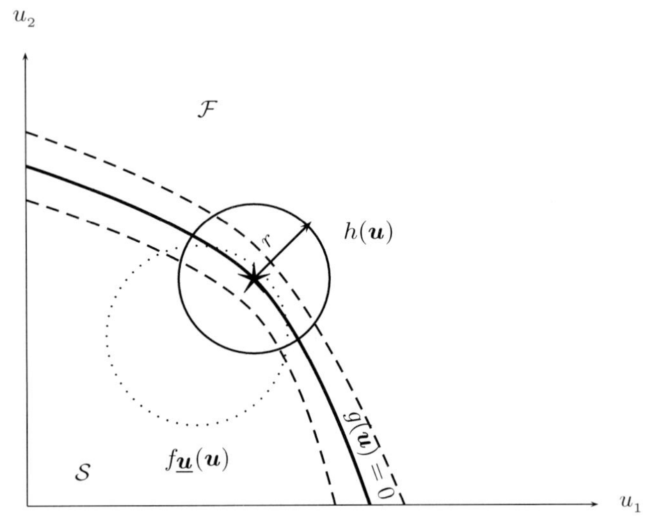

Fig. 5.7. Importance Sampling and Vapnik-Chervonenkis dimension.

5.7 Complexity, Dimensionality and Induction of SV Machines

Some important results of support vector classification are now presented as they are useful for a critical discussion on the usefulness of this approach for solving structural reliability problems. The first concerns the VC-dimension of optimal separating hyperplanes. It has been demonstrated [174] that it is bounded by

$$v \leq \min\left(\lceil \rho^2 ||\boldsymbol{w}||^2 \rceil, D\right) + 1 \,, \tag{5.32}$$

where ρ is the radius of the sphere containing the data in the feature space, $||\boldsymbol{w}||^2$ is the norm of the weight vector, D the dimensionality of the feature space and $\lceil z \rceil$ indicates the nearest integer to z such that $z < \lceil z \rceil$. Taking into account that while for any hyperplane of D dimensions $v = D + 1$ and hence it grows with the dimensionality with no bound, for optimal hyperplanes it is possible to overcome the curse of dimensionality if $\rho^2||\boldsymbol{w}||^2 < D$. In applying (5.32) it must be taken into account that the dimensionality of the feature space, D, is very high if the kernel satisfies Mercer's conditions, as shown in Chap. 3.

The dimensionality increase performed by the Support Vector methodology normally suffices for the inequality $\rho^2||\boldsymbol{w}||^2 < D$ to hold. However, in structural reliability applications it so happens that the radius (in input space) r is low in many cases, because the important region is well localized and hence a sampling density can be placed on top the design point for generating the training samples; this is the basic idea underlying the Importance Sampling concept (see Fig. 5.7) and the general algorithm for generating training samples exposed in Sect. 4.6. As a consequence, there is a high probability that most training samples are in a radius $r = 3$ if use is made of a standard Normal density in the standard space. It maps somehow in the feature space (see Sects. 5.8.6 and 7.5. A detailed discussion on the shape of the mapping is in [148]).

The above discussion indicates that the radius ρ of the training sample set in structural reliability is bounded. It remains to answer the question on the magnitude of weight vector in order to know whether $\rho^2||\boldsymbol{w}||^2 < D$ indeed holds even in the feature space. First, notice that the margin is given by (5.5): $\Delta = 1/||\boldsymbol{w}||^2$. Hence, a reduction of the margin implies an increase of the weight vector norm and vice-versa. One of the characteristics of structural reliability classification is that the classes are infinitely close to each other, as said in Chap. 2. Therefore, as the number of training samples inside the current margin increases, the margin decreases, with the consequence that if it becomes very large the product $\rho^2||\boldsymbol{w}||^2$ can become high. However, notice that (5.32) is valid in the feature space where the dimensionality D is very large. Also, an arbitrary reduction of the margin in a sequential approximation of the limit-state function would lead to overfitting, thus affecting the generalization capacity. For these reasons, it could be expected that in structural reliability applications the minimum in (5.32) corresponds to the term $\rho^2||\boldsymbol{w}||^2$. This will be demonstrated in a next example concerning the SORM approximation function proposed by [188].

Also important for the discussion of dimensionality issues of the SV methodology is the following bound for the generalization error P_{err} for a Support Vector Classifier trained with n samples [175]:

$$\mathrm{E}[P_{\mathrm{err}}] \leq \mathrm{E}\left[\min\left(\frac{s}{n}, \frac{\rho^2||\boldsymbol{w}||^2}{n}, \frac{D}{n}\right)\right] . \tag{5.33}$$

Here s is the number of support vectors. A naive approach to reduce the expected error probability is to increase the number of training samples n in order to reduce the third term. However, SVM rely on the other two terms in order to build good classification functions independently on the dimensionality. Even if $\rho^2||\boldsymbol{w}||^2$ is large, it suffices that s be less than D in order that the probability of classification error become independent of the number of dimensions of the optimization problem. This is frequently the case in image analysis applications, where values of the ratio s/d of the order of 10^{-3} and even lower have been found [175]. Therefore, the ratio s/D is much smaller. In structural reliability analysis applications, the number of support vectors can be made almost arbitrarily high by further and further refining of the Support Vector Classifier with sequential algorithms. This is because of the continuity of the safe and failure domains, which implies that it is always possible to find additional training samples in the current margin. According to the discussion on the curse of dimensionality done in Chap. 2, the higher the number of dimensions, the higher this probability. However such a refining is not necessary because of the very curse of dimensionality, since the classifier will be used for separating samples that will be very far from each other. In addition, the excessive refining of the classifier implies increasing its complexity, thus affecting its generalization capacity.

Finally, it is important to examine the application of the SRM inductive principle by support vector classifiers (see Chap. 2). Let us first recall the general form of the regularization principle (2.20), which is repeated here for convenience:

$$R_{\mathrm{r}}(\boldsymbol{\omega}) = R_{\mathrm{e}}(\boldsymbol{\omega}) + \lambda \Psi\left(h(\boldsymbol{x}, \boldsymbol{\omega})\right) . \tag{5.34}$$

First, note that in SVM the empirical risk $R_{\mathrm{e}}(\boldsymbol{\omega})$ is null if the classes are separable, as is the case in structural reliability because of the existence of an underlying performance function that is normally smooth. Hence, the generalization ability of the machines is enlarged if the complexity term $\Psi(\cdot)$ is minimized. According to (2.36) this can be achieved if the VC-dimension is minimized, which in turn can be attained as already discussed. However, instead of applying the regularization method, SVM automatically apply the Structured Risk Minimization principle summarized in Chap. 2 [175]. Let S_k be a set of hyperplane functions in the feature space satisfying

$$\rho^2 ||\boldsymbol{w}||^2 < k . \tag{5.35}$$

Hence, after (5.32), k is an upper bound to the VC-dimension because the dimensionality D of the feature space is very high. Since the SV classifier separates without error the training data in the feature space with a minimal weight vector norm (i.e. with maximal margin), it is concluded that it applies the SRM principle by its own procedures.

Another, practical regard on this issue is provided by the notion of sample complexity [4] summarized in Chap. 2. Recall that it is defined as the smallest integer that suffices for having a classification error e with a probability higher than $1 - \eta$. For $0 < \epsilon < 1/8$ and $0 < \eta < 1/100$ the following result applies:

$$N_{\mathsf{L}}(\epsilon, \eta) \geq \frac{v - 1}{32\epsilon} , \tag{5.36}$$

where v is the Vapnik-Chervonenkis dimension. This parameter is also bound from above by (5.32). Under the assumption that $\lceil \rho^2 ||\boldsymbol{w}||^2 \rceil < D$, as already justified, it is obtained that

$$N_{\mathsf{L}}(\epsilon, \eta) \geq \frac{1}{32\epsilon} \min \lceil \rho^2 ||\boldsymbol{w}||^2 \rceil . \tag{5.37}$$

This means that the number of training samples for Support Vector Machines can be selected regardless of the dimensionality of the input space.

In conclusion, the SV classification approach can overcome the curse of dimensionality in structural reliability if the following conditions are met:

- *The number of support vectors is less than the number of dimensions of the hyperplane.* Recall that this condition is essential for overcoming the curse of dimensionality by the formulation in the dual space of Lagrange multipliers, as said in Sect. 5.2.1. The dimensionality explosion caused by kernels satisfying the conditions of Mercer's theorem automatically induces a high dimensional space for which the condition $s < D$ is easily achieved.
- *The training data lie in a bounded sphere.* This condition is usually met in reliability problems because the training data are generated from a density function.
- *The norm of the weight vector is low.* Matching this condition in structural reliability analysis implies refraining from an excessive refinement of the classifier in order to avoid high complexity. The very curse of dimensionality makes this refining unnecessary as it has been noted. An example in the next section for a general SORM function shows that $||\boldsymbol{w}||^2$ decreases with D for a given number of training samples, thus facilitating the fulfillment of this condition.

As a consequence, since the above three conditions normally hold, and the first is specially powerful, the overcoming of the curse of dimensionality when using Support Vector Classifiers in structural reliability applications is assured. The following examples illustrate these issues as well as their practical implementation.

5.8 Application Examples

In this section the application of the SVM to solve structural reliability problems is illustrated with several examples.

5.8.1 Parabolic Limit State Function

In order to demonstrate the use of the basic algorithm exposed in Chap. 4, Figure 5.8 shows the decision function obtained by the support vector method for the parabolic limit state problem used in the comparisons made in that chapter. An initial population of 180 samples was generated in the surroundings of the known limit state function. In order to compress the data base, the 50 vector quantizers shown in the figure were calculated using the generalized Lloyd algorithm exposed in [27]. The ratio of quantizers to the population size, 50/180, was selected so large in this case for illustrating the good distribution of the quantizers. The support vector machine was calculated on the 15 quantizers that were the closest to the boundary. The probability of failure calculated with the resulting classifier was equal to the Monte Carlo simulation value.

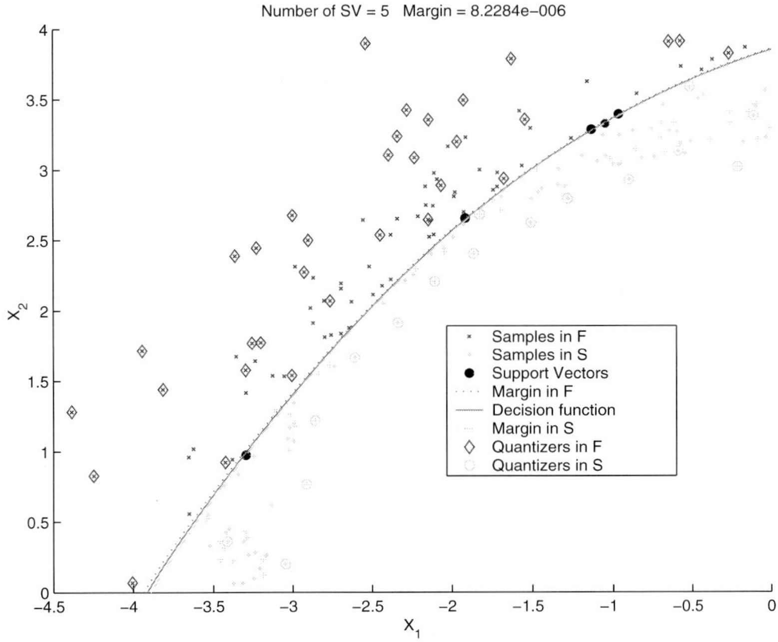

Fig. 5.8. Vector Quantizers and Support Vector approximation of a limit-state function

5.8.2 A Linear Limit State Function with Nonlinear Performance Function

An example concerning a two dimensional function studied by [96] is discussed next. We refer to the performance function

$$g(x_1, x_2) = \exp(0.2x_1 + 6.2) - \exp(0.47x_2 + 5)\ . \tag{5.38}$$

It is evident that this function crosses the basic variable space through a linear limit state despite its nonlinearity. Therefore, the example poses the following challenges:

- A regression approach in which a nonlinear model is specified a priori and is oriented to a global approximation (in other words, the RSM) would not capture the linearity of the class boundary.
- A local and flexible regression approach would be a better solution since the locality of the approximation would give the highest importance to

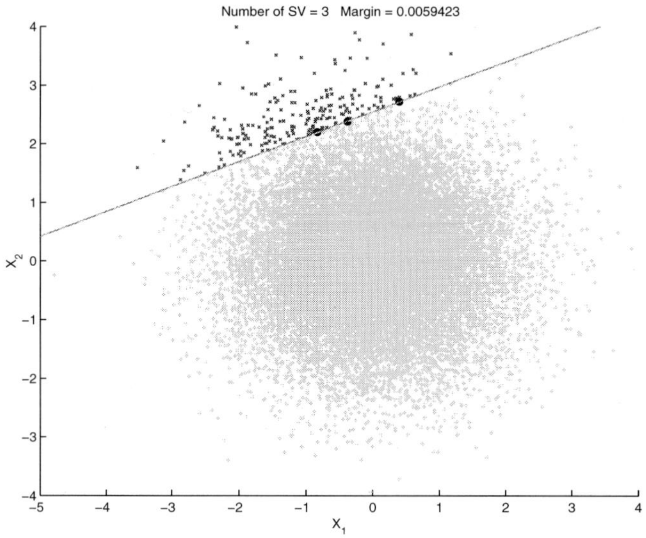

Fig. 5.9. On the advantages of flexible models for approximating limit state functions: Support Vector approximation of function (5.38) with a linear kernel

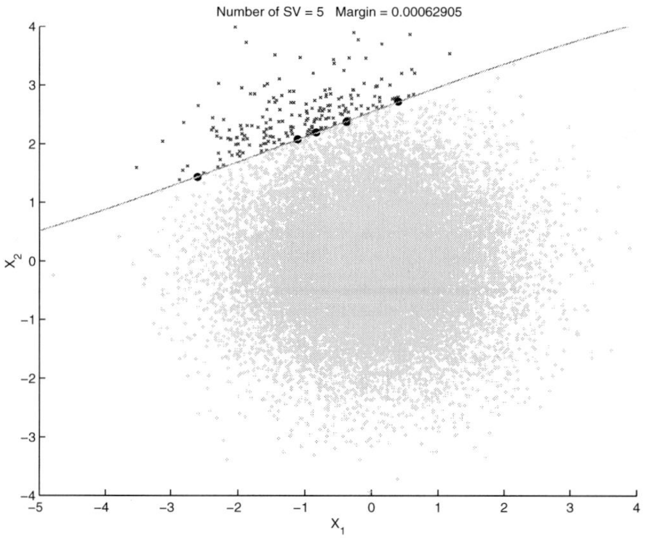

Fig. 5.10. On the advantages of of flexible models for approximating limit state functions: Support Vector approximation of function (5.38) with a radial basis kernel

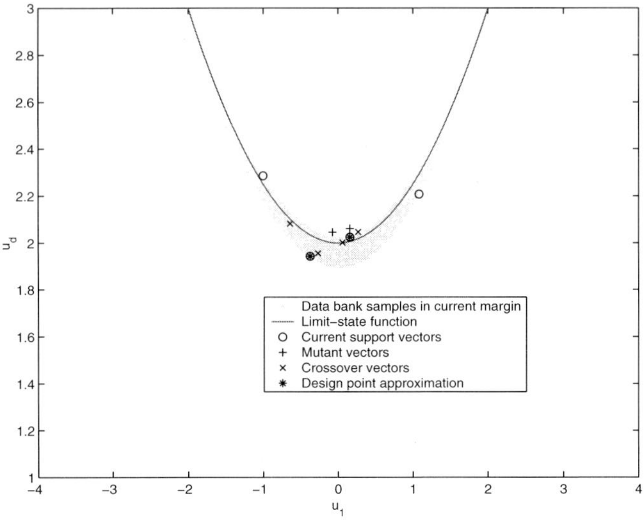

Fig. 5.11. Instances of convergence of the proposed SVM genetic algorithm for a two dimensional SORM problem

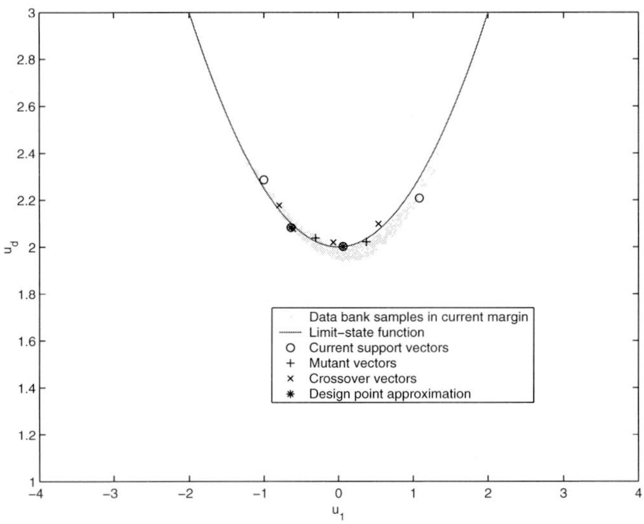

Fig. 5.11. *(contd.)*

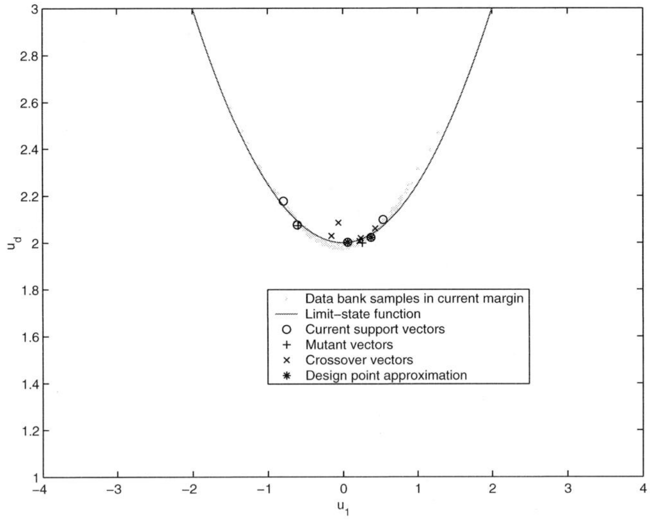

Fig. 5.11. *(contd.)*

samples in the vicinity of the boundary and the flexibility of the model allows its adaptation to the linearity of the limit state.

- A classification approach with such properties (like the SVM) would also be an excellent solution.

The last assertion is demonstrated by Figures 5.9 and 5.10, which have been obtained with the sequential algorithm based on margin shrinking. The former figure corresponds to the approximation with a linear kernel and the latter to a radial basis function kernel, which is highly nonlinear. In these approximations, the linear kernel required twenty calls of the function $g(\boldsymbol{x})$, three of which were selected by the optimization algorithm as support vectors, which are depicted in the figure with black circles. In the second case, the number of function calls is 26 and that of support vectors is 5. Notice that the three support vectors of the first approximation appear also in the second one, which illustrates a widely confirmed tendency of SVM, namely the relative kernel-independence of the approximation [148]. The figures also show a population of 20,000 classified test samples. Note that despite the second kernel is nonlinear, the flexibility and adaptivity of the model allow its good accommodation to the linear trace of the performance function in crossing the basic variable space. This means that it can be applied without a priori knowledge of the form of the limit state or the performance functions. Both classifiers were tested with 100,000 samples of Normal independent variates.

The error in the estimation of the probability of failure was null. In contrast, for the RSM the error is of one order of magnitude [96].

5.8.3 Two- and Twenty-Dimensional SORM Functions

In this example the instances of convergence of the proposed genetic algorithm for SVM training will be graphically illustrated with reference to the SORM approximating function (1.28), proposed by [188], which is repeated here:

$$u_d = \beta + \frac{1}{2R} \sum_{i=1}^{d-1} u_i^2 , \tag{5.39}$$

where R the average principal curvature radius. From this function the empirical second order index described in Chap. 7 was derived, which according to [188] yields better accuracy with respect to the exact results than other proposals if all principal curvatures have the same sign. According to the quoted reference all SORM methods yield very erroneous results if this is not the case. The function has been selected on the basis of the wide range of applicability of SORM formulas which, unfortunately, require the knowledge of gradients and principal curvatures that make their use very limited in practice.

The first analysis is of the two-dimensional case constituted by the set of parameters $(R, d, \beta) = (2, 2, 2)$. Notice that the curvature of the function is

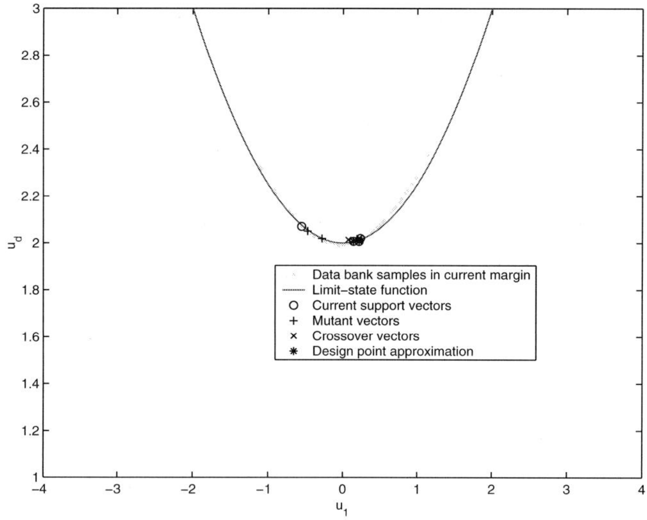

Fig. 5.11. *(contd.)*

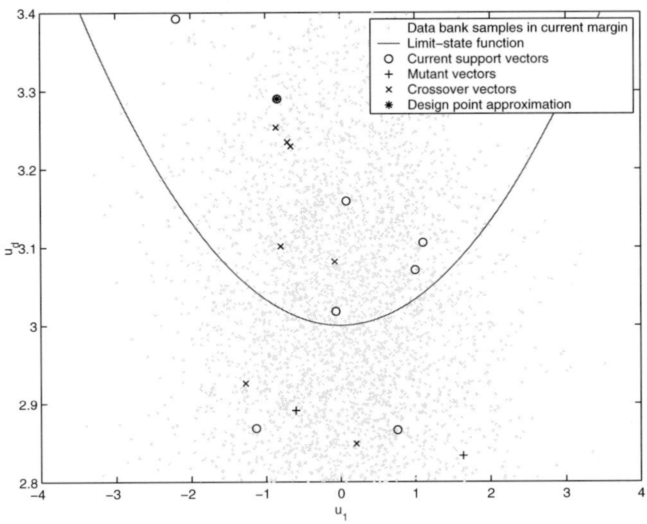

Fig. 5.12. Instances of convergence of the proposed SVM genetic algorithm for a twenty dimensional SORM problem

very high. A 10,000 sample pool was generated with a standard Normal density for all variables with the exception of the last one, which is a Normal displaced to the mean value $u_d = \beta$ and standard deviation of 0.2. The initial training samples were chosen from the data bank with the EM algorithm (see Chap. 3), as this assures the existence of samples in both the safe and unsafe domains. In applying the genetic algorithm a number of two mutants per iteration was specified. Figure 5.11 shows four instances of the convergence of the method to the limit-state function. The panels show the data bank samples that lie inside the current margin and the entire set of training samples accumulated to the current instance. Also appearing in the figures are the mutant and crossover vectors as well as the estimate of the design point. It can be observed that the margin is rapidly reduced and that the roles of the genetic mutation and crossover are as expected. The final approximations to the design point on both sides are practically coincident with the right value.

The proposed procedure is better tested with a larger dimensional case. The function used was the same as above with $(d, \beta) = (20, 3)$ and $R = 5\beta(d-1)$, which is far beyond the limit of applicability of FORM, according to [187]. Some instances of the approximation procedure in the (u_1, u_d) space are shown in Fig. 5.12. The meaning of the dots appearing in the figure is the same as above. Notice that due to the projection of the twenty dimensional space onto a bidimensional one some support vectors appear far from the limit-state function. The number of performance function evaluations was 181. The convergence to the design point is very rapid, as evident. The probability of failure obtained with the classifier and 50,000 samples was 0.0012, which coincides with the value obtained with the actual function and 500,000 samples.

Table 5.1. Frame example - Probability distributions of the basic variables

Variable	Distribution	Mean	Standard deviation
$\underline{E}_c$	Lognormal	1,960 kN/cm^2	196 kN/cm^2
$\underline{E}_b$	Lognormal	1,960 kN/cm^2	196 kN/cm^2
$\underline{I}_c$	Lognormal	100,000 cm^4	10,000 cm^4
$\underline{I}_b$	Lognormal	150,000 cm^4	15,000 cm^4
$\underline{P}_1$	Normal	24.50 kN	6.125 kN
$\underline{P}_2$	Normal	27.44 kN	6.860 kN
$\underline{P}_3$	Normal	28.42 kN	7.105 kN
$\underline{P}_4$	Normal	29.40 kN	7.350 kN
$\underline{P}_5$	Normal	30.38 kN	7.595 kN
P_6	Normal	31.36 kN	7.840 kN

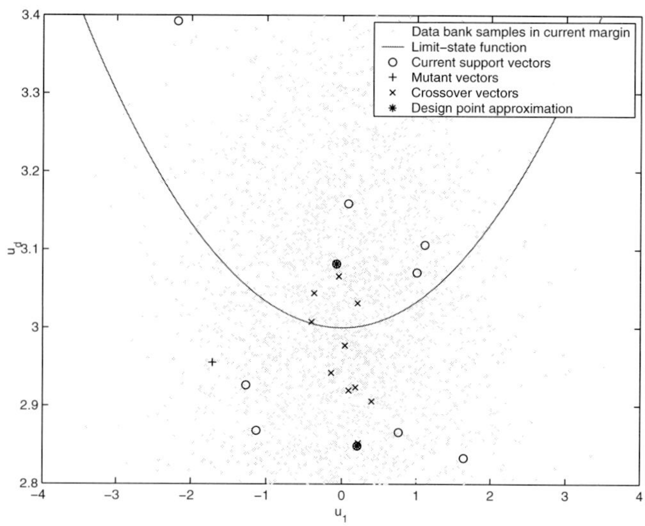

Fig. 5.12. *(contd.)*

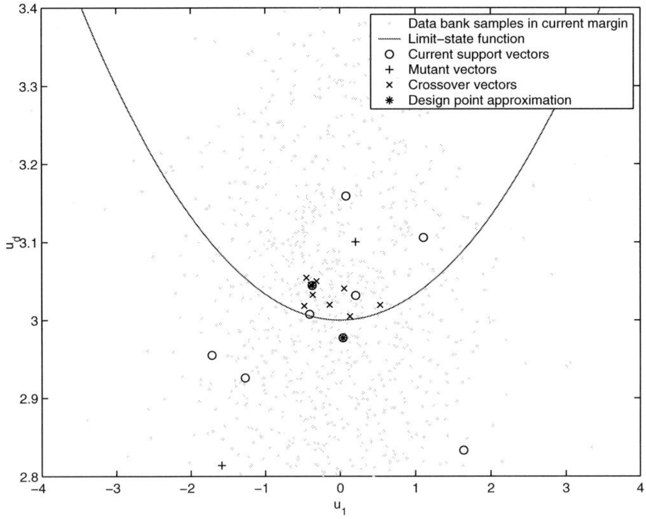

Fig. 5.12. *(contd.)*

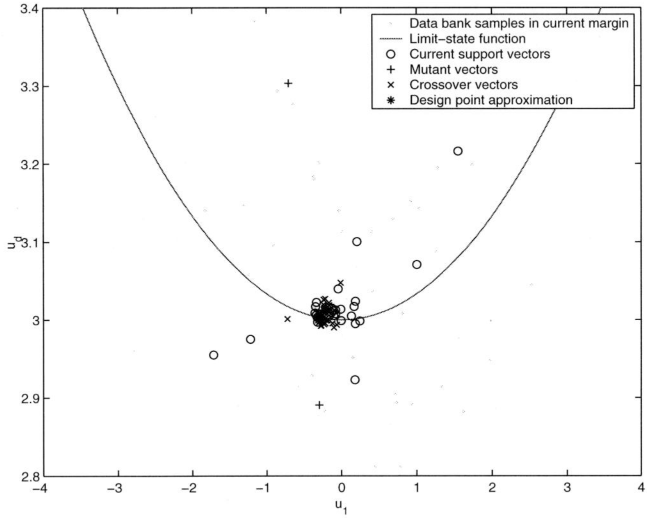

Fig. 5.12. *(contd.)*

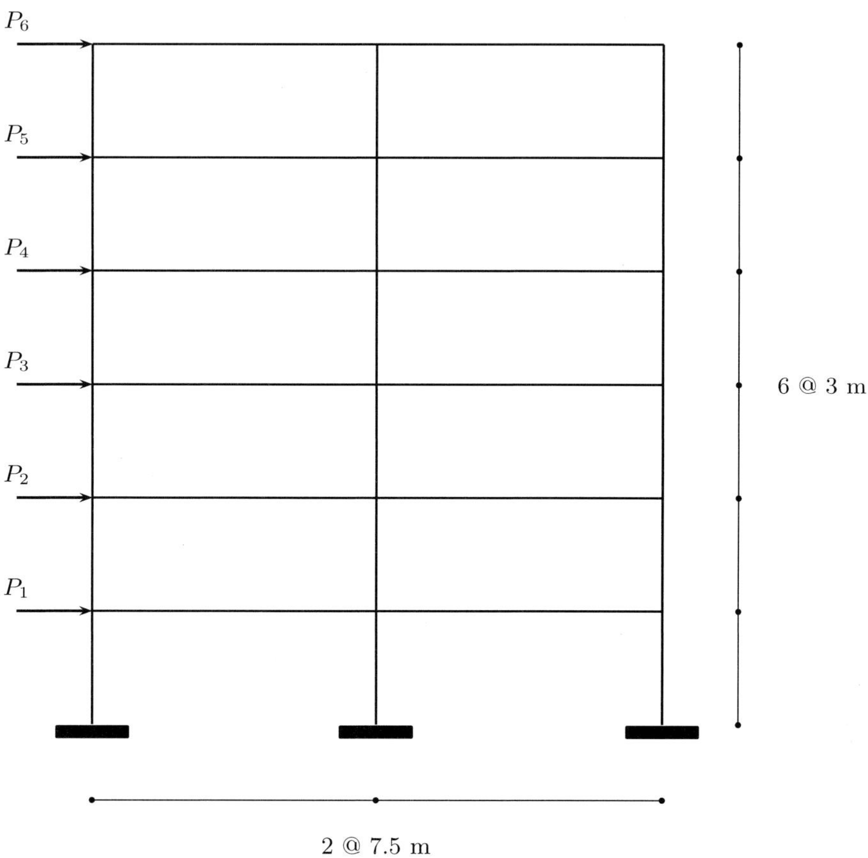

Fig. 5.13. Frame example

5.8.4 Ten Dimensional Problem

Consider now the linear frame shown in Fig. 5.13. The random variables are the modulus of elasticity E and the moment of inertia I of columns and beams $(E_{\mathrm{c}}, E_{\mathrm{b}}, I_{\mathrm{c}}, I_{\mathrm{b}})$ and the six horizontal loads, whose statistical description is shown in Table 5.1. The limit state function is now given by

$$g(\boldsymbol{x}) = 11\,\mathrm{cm} - r(\boldsymbol{x})\ , \tag{5.40}$$

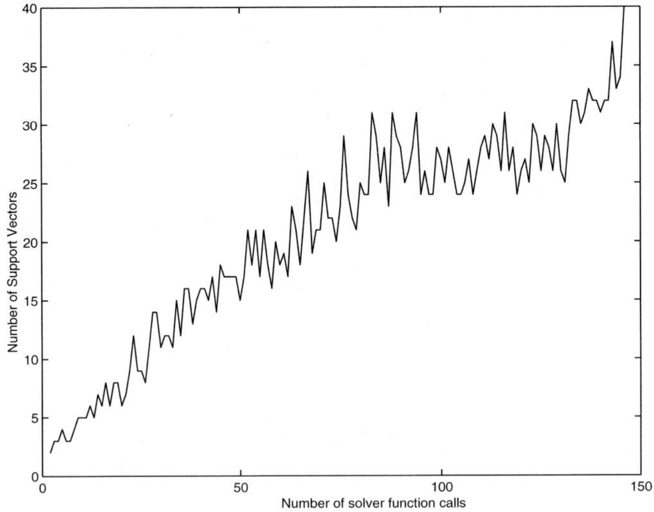

Fig. 5.14. Evolution of the number of support vectors

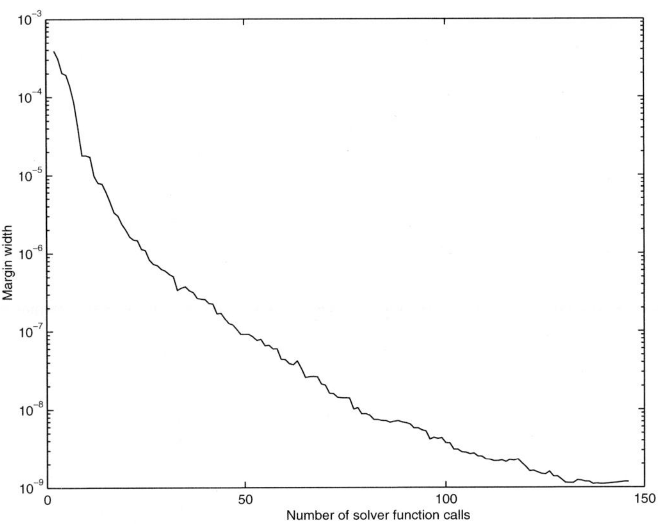

Fig. 5.15. Evolution of the margin

where $r(\boldsymbol{x})$ is the displacement at the right-top point of the frame, measured in centimeters. A probability of failure $\hat{P}_{\mathrm{f}_{\mathrm{MCS}}} = 5.451 \times 10^{-3}$ was calculated using 350,000 Monte Carlo simulations.

This case was evaluated with the radial basis function kernel. The algorithm based on margin shrinking reported in [80] and summarized above was started with only two samples, each one in a different safety domain. The SVM approach employed only 146 calls of $g(\boldsymbol{x})$. Once trained, the SVM was used as a solver substitute for the same data base employed in the Monte Carlo simulation. Only 5 samples out of the 350,000 were misclassified. The SVM-based probability of failure estimate was $\hat{P}_{\mathrm{f}_{\mathrm{SVM}}} = 5.457 \times 10^{-3}$. Therefore the relative error on the probability of failure estimate was 0.052% taking $\hat{P}_{\mathrm{f}_{\mathrm{MCS}}}$ as exact.

Figure 5.14 shows the evolution of the number of support vectors, while Figure 5.15 does the same for the margin. It can be observed that the number of support vectors tends to reach stationarity when the margin becomes very narrow and stable (notice the log-scale in Figure 5.15). It is evident that the training could have been stopped at an earlier time.

Let us define the efficiency of each solver call in a given method as $\hat{P}_{\mathrm{f}}/N_g$, where N_g is the number of $g(\boldsymbol{x})$ function calls. Then the higher the number of calls necessary to calculate a failure probability, the lower the efficiency of the method. The values of the efficiency ratio for the present case are the following: MCS: 1.559×10^{-8}; SVM: 3.738×10^{-5}. This means that the proposed approach is about 2,400 more efficient than standard Monte Carlo simulation in this example.

Before closing this example a discussion on the characteristics of the classification SVM approach is in order. First of all, the example shows that little information needed for starting the algorithm. This information can be obtained very easily. In fact, the two vectors needed for the algorithm to start can be obtained by simple setting two problems with non-conservative and conservative input data, respectively, selected at random. This is an important advantage of this approach over some techniques requiring, for instance, samples in the failure domain generated with the actual joint density function of the variables for initiating the procedure [3, 7], which is well neigh a small Monte Carlo simulation, or an assessment of the design point, like standard Importance Sampling. This is, in general, an advantage of solver substitution methods over techniques that do not uncouple the problem of the estimation of the limit state function from that of the calculation of the failure probability as such. When these problems are treated separately, the possibility of drastically reducing the number of function calls increases. The popularity of the RSM approach in the nineties confirms this assertion. In this uncoupling, the estimation of $g(\boldsymbol{x})$ can be managed with greater flexibility. However, the distinguishing advantage of the SVM is that its optimality is defined in terms of points close to the boundary, not on the mean square errors of the fitted polynomial, as in the RSM case. Moreover, only those vectors inside the current margin qualify as candidates to be the support vectors of the approx-

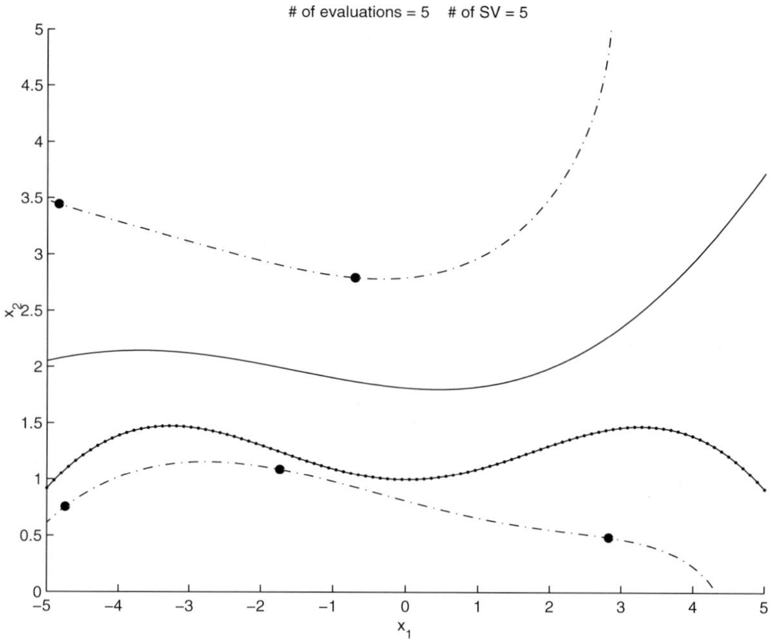

Fig. 5.16. Instances of convergence of the version space shrinking SVM algorithm for a two dimensional problem

imating boundary, making unnecessary the passing of all the sample points through the finite element solver.

5.8.5 An Application of the Version Space Algorithm

Let us consider the function

$$g(x_1, x_2) = 2 - x_2 - \exp\left(-\frac{x_1^2}{10}\right) - \left(\frac{x_1}{5}\right)^4 . \tag{5.41}$$

The pool was equipped with 25 unlabeled samples. Figure 5.16 shows three instances of the approximation to the function using the algorithm of version space shrinking. The good approximation properties of the method are evident.

5.8.6 Bound of the VC Dimension of the SORM Function

In order to examine the possibility of overcoming the curse of dimensionality with Support Vector Machines in the specific case of reliability problems, a numerical study was carried out in fitting an estimate of the SORM approximation function used in the previous example. Under the supposition that the condition on the equal sign of principal curvatures holds, the following numerical experiment was planned:

The target of the experiment is to determine the bound of the Vapnik-Chervonenkis dimension of the SVM approximation to (5.39) as a function of the dimensionality d. To this end, the average radius of curvature was fixed as $R = \beta(d-1)$, which is far from the applicability of FORM [187]. The value of d was varied in the range $[5, 120]$ and the value of the first-order reliability index β is 3.

A simple sequential procedure for approaching the limit-state function was adopted. It consists in learning the SVM with the same number of training samples, as justified by invoking the results on sample complexity. The 50

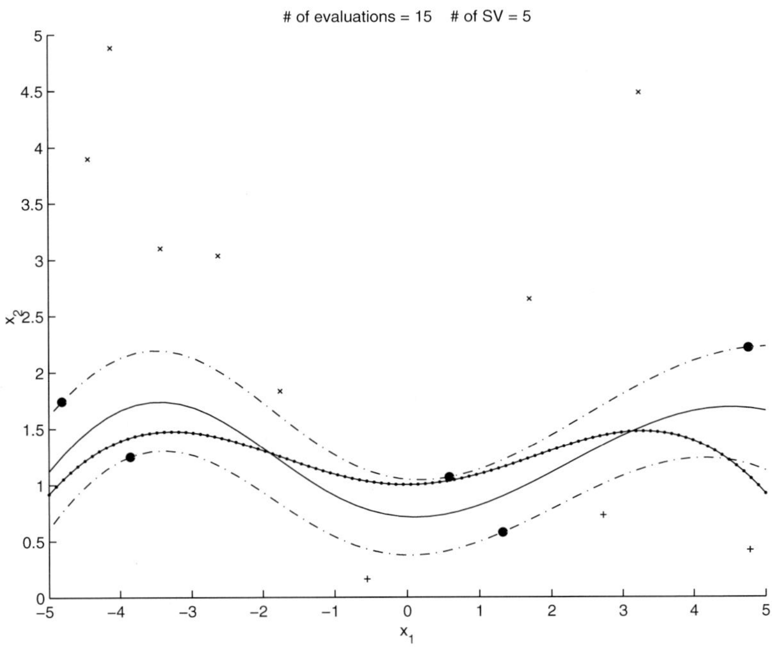

Fig. 5.16. *(contd.)*

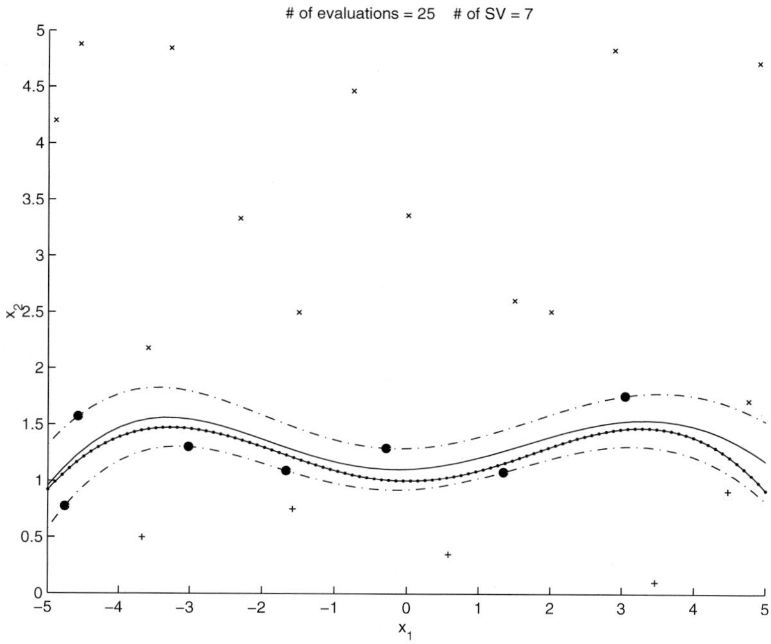

Fig. 5.16. *(contd.)*

training samples were applied in three steps. In the first step, ten random variables were produced with the Updated Latin Hypercube technique [49] and classified with the actual limit-state function. A trial SVM was then calculated with an homogeneous polynomial kernel (see Table 3.1) with $\theta = 1, p = 2$. The next forty samples were generated by the genetic algorithm described above. The value of the margin was obtained upon the end Support Vector Classifier using [174]

$$||\boldsymbol{w}|| = \left(\sum_{i=1}^{n} \alpha_i\right)^{\frac{1}{2}} . \tag{5.42}$$

For the inhomogeneous kernel the number of dimensions in feature space is given by (3.26), which is depicted in Figure 5.17. Using this kernel it is

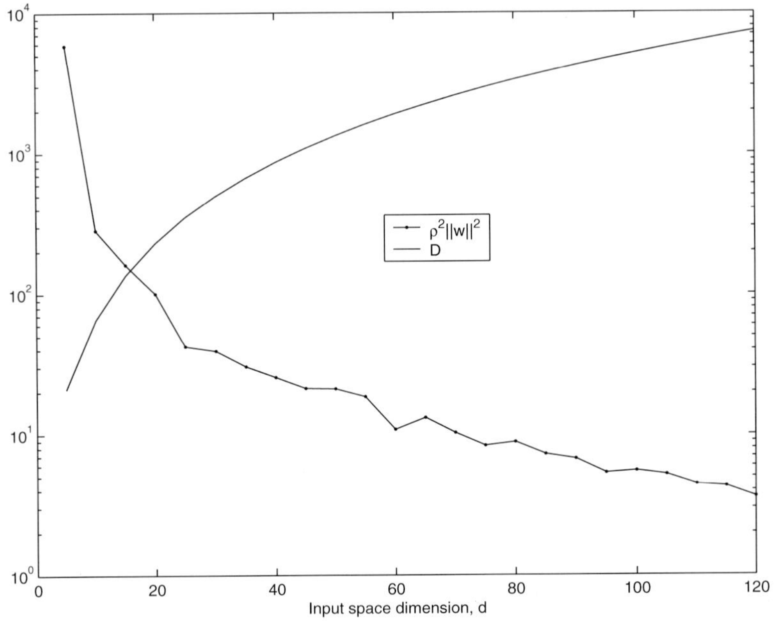

Fig. 5.17. Bound of the Vapnik-Chervonenkis dimension of a SORM function

possible to map any distance in input space q to the corresponding one in feature space Q and vice versa using

$$q = \sqrt{Q^{2/p} - \theta} \,. \tag{5.43}$$

Note that $q = 6$ is the diameter of a $3\sigma-$ball of Gaussian samples in the standard input space. Therefore the smallest ball in the feature space will have a radius $\rho \equiv Q = 37$. The results for $\rho^2||\boldsymbol{w}||^2$ are also shown in Fig. 5.17. It can be seen that for $d > 15$ the product $\rho^2||\boldsymbol{w}||^2$ becomes much less than the number of dimensions in feature space D (notice that the vertical scale is logarithmic). Notice that while $\rho^2||\boldsymbol{w}||^2$ rapidly decreases the dimensionality of the feature space does the opposite, thus assuring the overcoming of the curse of dimensionality for $d > 15$.

5.9 An Application to Stochastic Stability

The support vector genetic algorithm introduced previously can also be useful for alleviating the task of calculating the boundary between the stable and unstable domains, required in stochastic dynamic stability [102]. In this section we summarize this problem and then report the application of the algorithm.

5.9.1 Asymptotic Moment Stability

Let us consider a system of stochastic differential equations for a random vector $\underline{\boldsymbol{y}}(t)$ in the classical form [85]

$$\mathrm{d}\underline{\boldsymbol{y}}(t) = \boldsymbol{\Theta}(\underline{\boldsymbol{y}}(t), t)\mathrm{d}t + \boldsymbol{\Psi}(\underline{\boldsymbol{y}}(t), t)\mathrm{d}\underline{\boldsymbol{B}}(t) \ , \tag{5.44}$$

where $\boldsymbol{\Theta}$ is a $N-$dimensional vector of drift functions, $\boldsymbol{\Psi}$ is a $N \times R$ diffusion matrix and $\underline{\boldsymbol{B}}(t)$ is a vector of unit-strength, uncorrelated Wiener processes, that is,

$$\begin{aligned} \mathrm{E}[\mathrm{d}\underline{\boldsymbol{B}}(t)] &= \mathbf{0} \ , \\ \mathrm{E}[\mathrm{d}\underline{\boldsymbol{B}}(t)\mathrm{d}\underline{\boldsymbol{B}}^{\mathrm{T}}(t)] &= \boldsymbol{I}\mathrm{d}t \ , \end{aligned} \tag{5.45}$$

where $\boldsymbol{I}$ is the identity matrix. If the strength of the vector Wiener process $\underline{\boldsymbol{B}}(t)$ is different from the identity matrix, i.e.,

$$\mathrm{E}[\mathrm{d}\underline{\boldsymbol{B}}(t)\mathrm{d}\underline{\boldsymbol{B}}^{\mathrm{T}}(t)] = \boldsymbol{\Gamma}(t)\mathrm{d}t \ , \tag{5.46}$$

then matrix $\boldsymbol{\Psi}(\underline{\boldsymbol{y}}(t), t)$ can be modified to $\boldsymbol{\Psi}(\underline{\boldsymbol{y}}(t), t)\boldsymbol{\Gamma}^{\frac{1}{2}}(t)$ so as to produce a standard Wiener process as input. It can be shown that the solution of the Ito differential equation is a Markov process. An important tool for its solution is the Ito formula (e.g. [158]). For a differentiable function $h(y(t), t)$ it states that

$$\mathrm{d}h = \left[\frac{\partial h}{\partial t} + \theta\frac{\partial h}{\partial y} + \frac{1}{2}\psi^2\frac{\partial^2 h}{\partial y^2}\right]\mathrm{d}t + \psi\frac{\partial h}{\partial y}\mathrm{d}B \ , \tag{5.47}$$

which can be derived by expanding function $h(\cdot)$ in Taylor series up to second order terms, substituting the result into equation (5.44) and using the symbolic notation $\mathrm{d}B^2 = \mathrm{d}t$. Notice that the result differs from the one of ordinary calculus in the term $\mathrm{d}B$, which is $\sim \mathrm{d}t^{\frac{1}{2}}$ and hence cannot be neglected. In the general case of a function $h(\underline{\boldsymbol{y}}, t)$ of a $N-$ dimensional vector stochastic process $\underline{\boldsymbol{y}}(t)$ we have

$$\begin{aligned} \mathrm{d}h = \left[\frac{\partial h}{\partial t} + \sum_{i=1}^{N} \Theta_i(\boldsymbol{x}, t)\frac{\partial h}{\partial x_i} + \frac{1}{2}\sum_{r=1}^{R}\sum_{i,j=1}^{N} \Psi_{ir}(\boldsymbol{x}, t)\Psi_{jr}(\boldsymbol{x}, t)\frac{\partial^2 h}{\partial y_i \partial y_j}\right]\mathrm{d}t \\ + \sum_{r=1}^{R}\sum_{i=1}^{N} \Psi_{ir}(\boldsymbol{x}, t)\frac{\partial h}{\partial y_i}\mathrm{d}B_r \end{aligned} \tag{5.48}$$

If function $h(\cdot)$ is sequentially set equal to some given powers of the elements of $\underline{\boldsymbol{y}}(t)$, a system of stochastic differential equations can be obtained after replacement of these functions into the Ito formula. This system can then be converted to a system of ordinary differential equations (ODE) describing the moment evolution by applying the expectation operator to the result and taking into account that $\mathrm{E}[\mathrm{d}B_r] = 0$ as said in the foregoing. To this purpose function $h(\underline{\boldsymbol{y}})$ is set equal to

$$h(\underline{\boldsymbol{y}}) = y_1^{j_1} y_2^{j_2} \dots y_N^{j_N} \ . \tag{5.49}$$

The exponents $j_i, i = 1, 2, \dots, N$ can be considered as elements of the multi-index vector $\boldsymbol{j} = [j_1, j_2, \dots, j_N]$ whose sum of elements $|\boldsymbol{j}| = j_1 + j_2 + \dots + j_N$ indicates the order of the moment being evaluated. Taking into account that $\mathrm{E}[\mathrm{d}\underline{\boldsymbol{B}}(t)] = \boldsymbol{0}$ and that function $h(\cdot)$ is not an explicit function of time, it can be easily shown that the application of the expectation operator to the Ito formula yields the following sets of moment equations:

$$\dot{\boldsymbol{\mu}}_J = \boldsymbol{g}_J[\boldsymbol{\mu}_1, \boldsymbol{\mu}_2, \dots], \qquad J = 1, 2, \dots \ . \tag{5.50}$$

Here $\boldsymbol{\mu}_J$ denotes the vector of all moments $\mu_{\boldsymbol{j}}$ for which $|\boldsymbol{j}| = J$ and $\boldsymbol{g}_J$ are vector linear functions. In most practical cases the ODE system (5.50)is linear in the moments so that it can be written in the form

$$\dot{\boldsymbol{\mu}}_J = \boldsymbol{A}_J \boldsymbol{\mu}_J, \qquad J = 1, 2, \dots \ . \tag{5.51}$$

The asymptotic moment stability of the structure is assured if the eigenvalues of matrices $\boldsymbol{A}_J$ have a negative real part [102]. However, the computation all these eigenvalues in a trial and error procedure for finding the boundary between the stable and unstable domains can be a time consuming task, which is complicated by the asymmetric structure of the matrices. A further complication stems from the fact that the number of equations grows rapidly with the number of degrees of freedom and the order of the moments J. In fact, for a state vector of length N and the order of moments equal to J, the number of moment equations is

$$M = \binom{J+N-1}{J} = \frac{(J+N-1)!}{J!(N-1)!} \ , \tag{5.52}$$

as this is the equation for placing J objects into N cells, which can be empty, without distinguishing the balls [165]. Hence for $N = 10$ (five degrees of freedom) we have for $J = 2$, $M = 55$ and for $J = 3$, $M = 220$. These figures illustrate the convenience of having a numerical procedure that minimizes the number of eigenvalue or determinant evaluations.

5.9.2 Numerical Example

Consider the bridge section shown in Fig. 5.18, in which the degree of freedom of interest is the torsional rotation α. The equation of motion is [156]

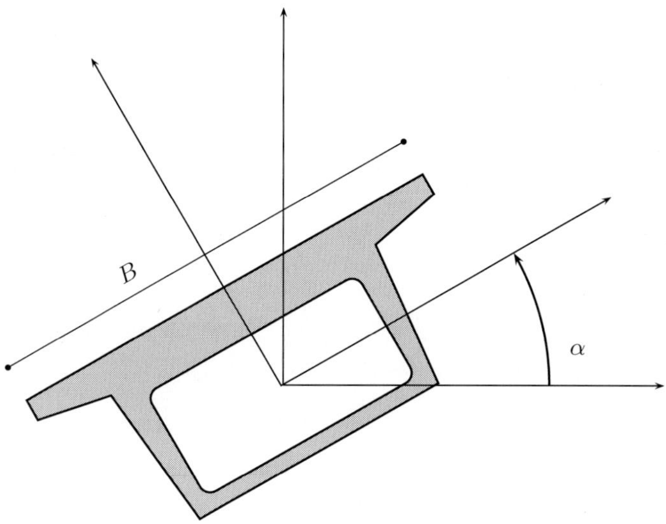

Fig. 5.18. Torsional stability of a bridge under wind loads

$$I\left(\ddot{\alpha} + 2\xi\omega\dot{\alpha} + \omega^2\ddot{\alpha}\right) = \frac{1}{2}\rho(2B^2)p_1$$
$$\times \left[p_2 w^2(t)\dot{\alpha} + w^2(t)\alpha + \int_{-\infty}^{t} w^2(t)\left(f_\alpha(t-\tau) - 1\right)\frac{\partial\alpha(\tau)}{\partial\tau}\mathrm{d}\tau \right] , \quad (5.53)$$

where I is the moment of inertia per unit length, ω is the natural frequency, ξ the damping ratio, ρ the air density, B the width of the bridge deck, w the wind velocity and $f_\alpha(t)$ the so-called indicial function. In this example we will start from the conversion of this ordinary differential equation into a system of stochastic differential equations exposed in detail in [85], under the following assumptions:

- The wind velocity is a stochastic process $W(t) = \mu_W(1+N(t))$, where μ_W is the mean velocity and $N(t)$ a random process, whose strength is low. This allows writing $W^2(t) \approx \mu_W^2(1+2N(t))$.
- The indicial function can be approximated as

$$f_\alpha(t) \doteq 1 + c_1 \exp(-\gamma_1 t) + c_2 \exp(-\gamma_2 t) , \quad (5.54)$$

where c_1, c_2, γ_1 and γ_2 are empirical parameters.

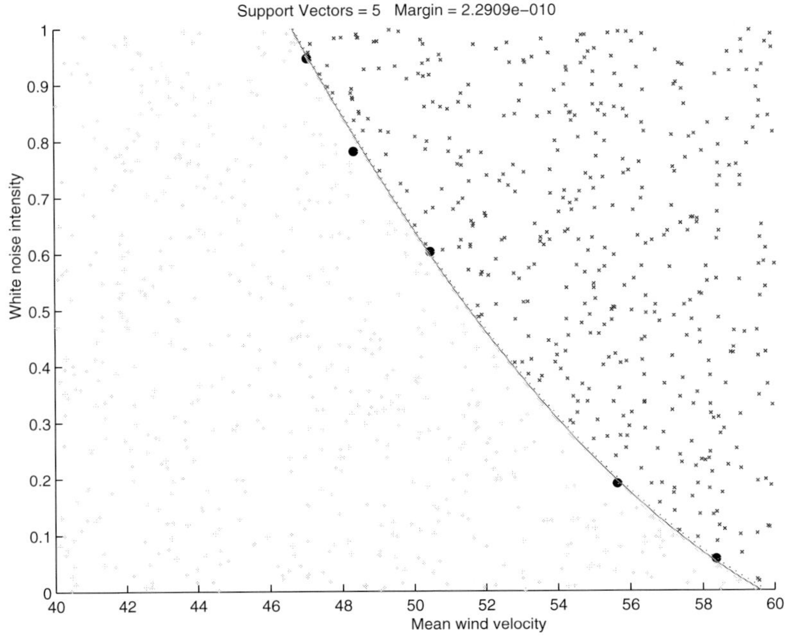

Fig. 5.19. Stochastic stability boundary of the bridge

- The velocity random process is wide-band and hence it can be transformed in a white noise of intensity D_0 using the Wong-Zakai correction terms [158].

After applying these hypotheses the following system is arrived at:

$$\mathrm{d}\underline{\boldsymbol{y}} = \boldsymbol{\Theta}\mathrm{d}t + \boldsymbol{\Psi}\mathrm{d}\underline{\boldsymbol{B}}(t)\ , \tag{5.55}$$

where $\boldsymbol{y} = [\alpha\ \ \dot{\alpha}]^{\mathrm{T}}$,

$$\boldsymbol{\Theta} = \begin{pmatrix} 0 & 1 & 0 & 0 \\ -\omega_0(\omega_0 - 2a_1a_2D_0) & -2\omega_0(\xi_0 - a_2^2a_2\omega_0D_0) & a_1 & a_1 \\ 2c_1a_1D_0 & c_1(1+2a_2\omega_0D_0) & -\gamma_1 & 0 \\ 2c_2a_1D_0 & c_2(1+2a_2\omega_0D_0) & 0 & -\gamma_2 \end{pmatrix}, \tag{5.56}$$

and

$$\boldsymbol{\Psi} = 2 \begin{pmatrix} 0 \\ a_1 y_1 + a_2 \omega_0 y_2 \\ c_1 y_2 \\ c_2 y_1 \end{pmatrix} \sqrt{D_0} \ . \tag{5.57}$$

The data for the analysis have been drawn form [161]:

$$\begin{array}{rrr} B = 28\text{m} & \xi = 0.01 & \omega = 0.74\pi \ , \\ I = 1.53 \times 10^6 \text{m}^4 & \rho = 1.266 & p_1 = 0.28954 \ , \\ c_1 = 0.13153 & c_2 = 1.13159 & g_1 = 1.08272\mu_W / B \ , \\ g_2 = 1.06556\mu_W / B & a_1 = \rho \mu_W^2 & B^2 p_1 / I \ , \\ \omega_0 = \sqrt{\omega^2 - a_1} & a_2 = -0.14346\rho\mu_W B^3/(I\omega_0) & p_2 = a_2 \omega_0 / a_1 \ , \\ \xi_0 = (2\omega\xi - a_1 p_2)/(2\omega_0) \ . & & \end{array} \tag{5.58}$$

The application of the Ito formula leads to the following equation for estimating the stability boundary in second moment sense, which is stronger than that of the first moment:

$$\dot{\boldsymbol{\mu}}_2 = \boldsymbol{A}\boldsymbol{\mu}_2 \ , \tag{5.59}$$

where

$$\begin{aligned} \boldsymbol{\mu}_2 = [&\mu_2(2,0,0,0)\ \mu_2(0,2,0,0)\ \mu_2(0,0,2,0)\ \mu_2(0,0,0,2)\ \mu_2(1,1,0,0)\ , \\ &\mu_2(1,0,1,0)\ \mu_2(1,0,0,1)\ \mu_2(0,1,1,0)\ \mu_2(0,1,0,1)\ \mu_2(0,0,1,1)]^{\mathrm{T}} \ . \end{aligned}$$

The details of the matrix $\boldsymbol{A}$ can be read in [161].

Conventionally, the computation of the stability boundary in the (μ_W, D_0) space is carried out by fixing a noise intensity and then increasing the mean velocity until a change of sign of any eigenvalue occurs [23, 161]. Since this is a trial and error procedure it requires many eigenvalue solutions. In contrast the application of the proposed procedure implied only 32 evaluations of the determinants of matrix $\boldsymbol{A}$. Figure 5.19 shows the approximating boundary, which comprehends five support vectors appearing as black circles. In order to test the accuracy of the classifier, one thousand samples were generated and classified with the support vector function as shown in the figure (The + and × signs denote stable and unstable samples, respectively). It is evident that the accuracy of the method is very good. If a less accurate classifier suffices for practical purposes, the sequential procedure can be stopped at an earlier phase, implying less computations of the eigenvalues of the system matrix.

Similar applications of the classification techniques for stochastic stability analysis can be made in other fields, such as that of structural control, in which the reliability is similarly determined by the negativity of the real part of the state matrix eigenvalues [160] .

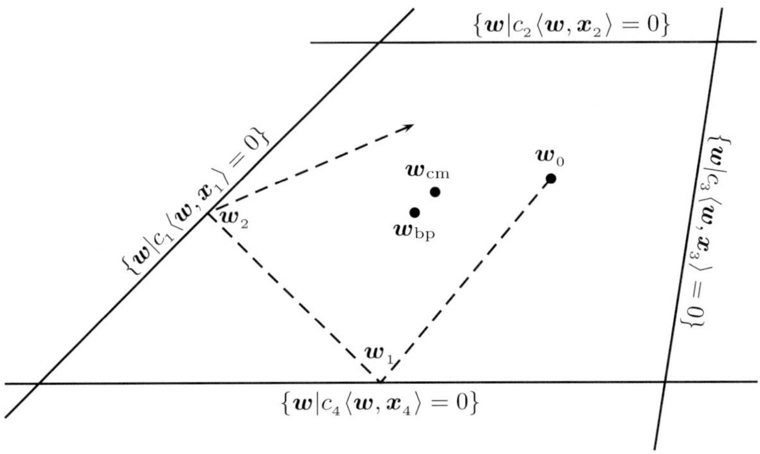

Fig. 5.20. Kernel billiard algorithm for Bayes Point Machines. The optimal Bayesian solution $\boldsymbol{w}_{\text{bp}}$ is approximated by the center of mass of the version space $\boldsymbol{w}_{\text{cm}}$. This is estimated by a billiard algorithm: Starting from an initial guess $\boldsymbol{w}_0$ the ball bounces at points $\boldsymbol{w}_i, i = 1, 2, \ldots$ and the estimate is their statistical average. (Adapted form [70]

5.10 Other Kernel Classification Algorithms

In this section some alternatives to the SV method for classification are briefly discussed. Most of these alternatives can be grouped under the name of *Bayesian kernel methods*, as they implement a combination of the Bayes theorem and the kernel trick in different ways. For an in-depth treatment of the two selected Bayesian techniques (Bayes Point Machines and Relevance Vector Machines) and of other Bayesian kernel approaches (Gaussian Processes, Laplacian Processes, etc) the reader is referred to [70] and [148].

- *Bayes Point Machines* [71]. This algorithm is based on the concept of version space. By formulating the hypothesis of a uniform prior distribution of the weights on this space, a Bayesian estimate of the class sign corresponding to a sample vector is derived. This is called the *Bayes classification strategy*. The *Bayes point* is defined as the weight vector that minimizes the expected value of zero-one cost function (see 2.15) that counts the discrepancies between the Bayes classification strategy and the classification given by the hypothesis hyperplanes lying in the version space. Since this computation is difficult, it is replaced by a surrogate, namely the center of mass of the version space, for the estimation of which a Markov chain Monte Carlo method called *kernel billiard* has been proposed (see Figure 5.20). The kernel trick is introduced at a convenient step in the algorithm.

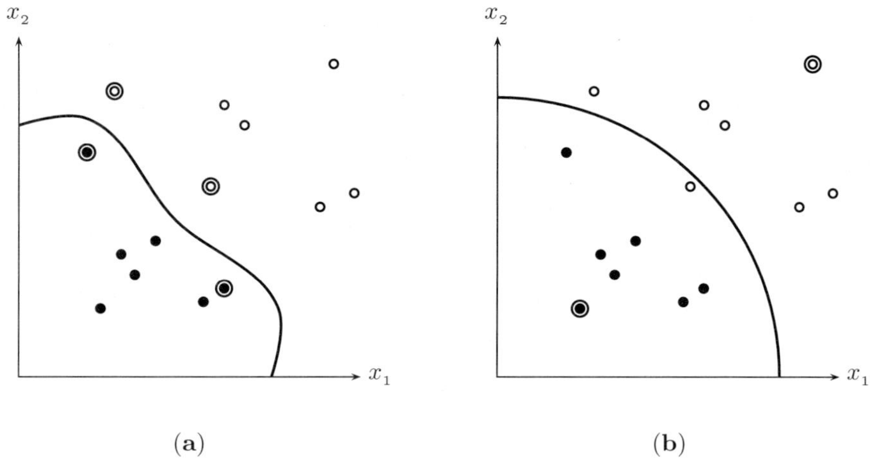

Fig. 5.21. Comparison of Relevance and Support Vector Machines. SVM (left) are determined by samples close to the boundary whereas RVM (right) by samples far from it. In non separable problems the number of RV tends to be lower than that of SV and hence the surface they determine is normally smoother at the expense of sacrificing the relevance of samples close to the boundary

- *Relevance Vector Machines* [168, 169]. This algorithm is based on the maximum likelihood of the posterior probability of the class assignments given an hypothesis of the parameters governing the model. These latter are not only the weights of the linear decision function (which has the same form as that of the SVM, i.e. by means of kernel functions) but also those of the Gaussian prior of the class labels. As a result, the solution for the decision function depends on some relevance vectors, whose number is normally lower than the number of SV, and which are located far from the boundary (Fig. 5.21).
  With respect to these differences between SVM and RVM two comments are in order: (a) The reduction in the number of representative vectors and hence of kernel evaluations, which is stressed by the creators of the RVM

technique as its main feature, tends to be more accentuated in non separable problems than in separable ones, to which category the structural reliability problem belongs. (**b**) The overcoming of the dependence of the boundary in RVM and the assumption of underlying Gaussian densities suggest that the method would not work well for small samples sets surrounding the limit state function, which is the ideal in structural reliability applications.

- *Other algorithms.* The *Least-Square Support Vector Machines* [163] are basically a modification of the SVM for non-separable classes (see Chap. 6, Sect. 6.4). For perfect separability the method coincides with the standard SVM and therefore it is of little interest in the structural reliability context. Something similar occurs with the *Adaptive Margin Machines* [70] , which is an algorithm that only makes sense in the non separable case.

6

Regression Methods

6.1 Introduction

After having examined the classification approaches to reliability problems in the previous two chapters, we will be concerned herein with the applicability of function approximation methods, with special regard to Neural Networks and Support Vector Machines. Three kinds of networks are of interest in this respect: Multi-Layer Perceptrons, Radial Basis Function Networks and Networks with Time-Dependent connections. The use of Support Vector Machines for regression purposes is based on the extension of the method to non-perfectly separable classes.

As said in Chap. 1, the construction of a solver-surrogate under a regression approach can be used either in connection to methods based on the limit-state function or on the performance function, while classification methods can only be applied in the former case. Most of the (few) uses of Neural Networks reported in the structural reliability literature correspond to the former viewpoint. For this reason, the later will be stressed in this chapter, as it allows the estimation of the entire distribution function of the observed responses, which is of general interest in structural reliability.

6.2 The Response Surface Method Revisited

Despite not developed in the field Statistical Learning tool but in that of Design of Experiments, the Response Surface Method is also a regression approach. Its limitations for structural reliability applications have been underlined in Chap. 1. However, some further comments on the method can be made in this chapter on the basis of the Statistical Learning concepts summarized in Chap. 2.

6.2.1 Dimensionality Problems

In Chap. 1 it was indicated that the RSM requires the repetition of the experimental plan R times for reaching an approximation of the performance function in the surroundings of the design point (see Fig. 1.13). For $d = 9, R = 3$, 57 and 1593 function evaluations are needed when using the Saturated or the Central Composite Designs, respectively. Accordingly, the dimensionality is specially serious when using the latter design, which is reputed to be the most accurate option [63].

Let us now turn the attention to the VC-dimension. According to the theory of statistical learning, the Vapnik-Chervonenkis dimension of a linear model like

$$\sum_{i}^{m} w_i h_i(\boldsymbol{x}) \tag{6.1}$$

is $v = m+1$ [27]. Note that in this model the functions $h_i(\boldsymbol{x})$ are non-adaptive to the available samples and have no parameters that are the object of the training. These conditions are met by the power functions employed in the RSM, as said in Chap. 1. According to Table 1.2 this means that the VC-dimension of the RSM models is $v = 2d + 2$ for the Saturated Design and $v = 2^d + 2d + 2$ for the Central Composite Design. For this reason, the error bound of the generalization regression risk (2.40) increases either linearly or geometrically with the dimensionality.

Such dependence of the VC-dimension on the dimensionality also occurs in multi-layer perceptrons, and is in contrast to the independence that characterizes the support vector methodology, as already said in Chap. 5. However, as shown in Sect. 4.4.3, multi-layer perceptrons can effectively overcome the curse of dimensionality in practice, as the error tends to decrease with increasing d, while the polynomial fitting applied by the RSM shows the opposite trend, as shown in Fig. 4.18, which corresponds to a second-degree polynomial. If, on the other hand, such degree is raised, (4.40) indicates that the training (i.e. bias) error decreases but the generalization (i.e. variance) error increases, as the complexity of the model raises in parallel.

6.2.2 Performance Function Approximation

The distinction made in this book between the performance and limit state functions points out another contrast between the flexible classification methods and the traditional RSM for regression. This concerns the training sampling needed in both cases. In fact, for approximating the performance function $g(\boldsymbol{x})$ it is necessary to scan the function in a domain like that shown in Figure 1.13, in order to reduce the risk of error in a wide region of the input variable space. Since this technique is oriented by a regression approach intended to approximate the performance function, it should perform some iterations by moving the experimental plan towards the limit-state function. On the contrary, for a direct approximation of the function a sampling on both sides of the function and very close to it, as illustrated by Figure 1.23, is desirable. Moreover, since the actual use of the approximation to the performance function is as a classifier, the research efforts in structural reliability should be addressed to this kind of sampling.

The orientation of the RSM towards the performance function is a consequence of the fact that this technique is borrowed from the field of Design of Experiments, in which the central matter is the effect of the variables on the system's response and not the classification of samples. This points out another contradiction of the conventional RSM approach that was understood by [96], who clearly state that "to improve the response surface method, the fitting points should be positioned closely to the original limit state" [96], as in Figure 1.23. Accordingly, they propose a method called Vector Projected Points that, from the viewpoint of Statistical Learning Theory, can be considered to be in the direction to the support vector methodology in the sense that it is aimed to locate samples as close as possible to the unknown limit state function, disregarding the values of the performance function far from the class boundary.

6.2.3 Naive Inductive Principle

In the same spirit, the application of the Response Surface Method normally consists in a simple application of the naive ERM principle over a model defined a priori for approximating the performance function. If the number of solver calls equals that of model coefficients, the empirical risk is null. In spite of this, such a model imposition can lead to model suboptimality. This phenomenon has been observed by Kim and Na in the above quoted paper [96]. The issue of the suboptimality of this rigid model is illustrated by these authors with the following example, already considered in the previous chapter (5.38):

$$g(x_1, x_2) = \exp(0.2x_1 + 6.2) - \exp(0.47x_2 + 5) . \tag{6.2}$$

As said before, this function happens to form a linear limit state function in the space (x_1, x_2). However, the nonlinearity of the performance function suggests the adoption of a quadratic polynomial for approximating it. This leads to errors of about one hundred percent in the estimation of the failure probability [96]. The example is interesting in that the performance function approach to simple reliability analysis (i.e. the estimation of a single probability) bears a contradiction, because a model that seems adequate for the performance function $g(\boldsymbol{x})$ is thoroughly inadequate for the limit-state function $g(\boldsymbol{x}) = 0$, which is the real target of the analysis. This supports the use of different names for these functions adopted in this book. More importantly, note that, since the RSM applies the simple ERM inductive principle, it cannot automatically penalize the use of a more complex (or a more simple) model than needed, as is commonly done in using neural networks with regularization techniques such as weight decay, early stopping, etc, mentioned in Chap. 1, or in using Support Vector Machines which automatically apply the SRM principle. As a consequence, if the RSM model specified by the analyst is not the optimal one, large generalization errors could be expected, as in the just quoted example. The problem is that the analyst cannot know this beforehand. Finally, despite the cute observations made by [96] on the conventional RSM, the method they propose (a linear response surface) can be suboptimal in some cases by the same token.

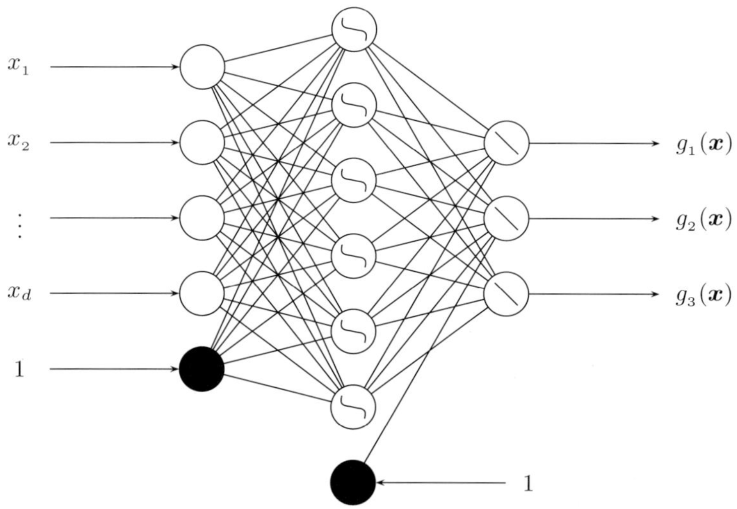

Fig. 6.1. Two-Layer Perceptron useful for reliability analysis treated as a regression task

6.3 Neural Networks

Neural networks of the MLP and RBFN types have been applied in recent years to performance function-based reliability analysis [26, 123, 144, 79, 77, 145, 25, 59]. The dominant trend in these applications is on the basis of some previous information on the design point and in combination with variance-reduction sampling techniques. Figure 6.1 shows a two-layer perceptron that is useful for reliability of full probabilistic analyses under the regression approach. It comprises a hidden layer of sigmoid functions and an output layer with linear ones. The three neurons in this layer indicate that the same per-

Table 6.1. Frame example in Figure 5.13

No. of hidden neurons	$\hat{P}_{\mathrm{f}}^{*}$
6	0.0053
8	0.0053
6	0.0054
7	0.0053
6	0.0053
5	0.0053
4	0.0054

ceptron could be used for estimating simultaneously the responses for the same number of structural responses.

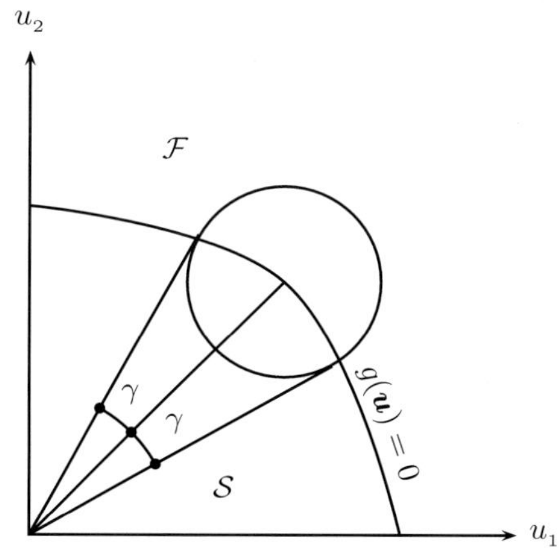

Fig. 6.2. Directional simulation strategy used for training MLP

As an example, consider the frame illustrated in Fig. 5.13 which was solved with Support Vector classifiers in the previous chapter. The problem was also solved with Multi-Layer Perceptrons trained with samples generated by means of a directional simulation technique [77]. First, the location of the design point was estimated using a preliminary sample set. Then the two-layer networks were learned with 150 samples generated in the following manner: The training samples are selected as those lying in the union of the hypersphere placed around of the estimated design point and its corresponding hypercone defined as

$$\cos \gamma = \frac{\beta}{\sqrt{\beta^2 + \kappa^2}} \tag{6.3}$$

in the standard $\boldsymbol{u}-$space (see Fig. 6.2). The parameter κ is the radius of the hypershere. For the analysis a value of $\kappa = 3$ was selected. Table 6.1 displays the results of the failure probability estimate for different architectures of the perceptrons. Taking into account that the exact probability is 0.0054, it is evident that this approach is accurate as well as stable in terms of the network structure.

Since the use of neural networks in this sense has been the dominant trend in structural reliability applications in recent years, as evidenced in the quoted papers, this issue will not be illustrated further herein. More relevant are (a) advanced forms of their use such as boosting and (b) their use for a full probabilistic analysis. These important issues are examined in the sequel.

6.3.1 Boosting

In recent years, part of the research in pattern recognition has been addressed to the use of statistical resampling techniques such as the jackknife or the bootstrap (see e.g. [109]) or to the combination of several classifiers to produce an improved one. In the first approach we have the so-called *bagging algorithms* (for bootstrap aggregation) [19]. When applied to Neural Networks, this method is known as *committee of networks*. There are several methods to build such a committee. One of the most powerful is the technique known as *boosting* in which a series of classifiers is built for learning only those samples that the previous ones have found to be the most difficult to classify. The final model is a weighted linear combination of all of them, in such a way that the weight of each model is determined by its failure to classify correctly the samples it is trained with. Formally we have the following classifier:

$$c(\boldsymbol{x}) = \operatorname{sgn}\left[\sum_{t=1}^{T} \gamma_t c_t(\boldsymbol{x})\right] , \tag{6.4}$$

where $c_t(\boldsymbol{x})$ is a classifier component, γ_t its weight (also called *vote*) and T the maximum number of components, which can be regarded as the number of iterations from the algorithmic point of view.

Despite its apparent simplicity, boosting algorithms have deep roots in statistical learning principles. This is especially true for the AdaBoost algorithm (for "adapting boosting")[50] which may yield a classifier of arbitrary accuracy with respect to the training set, as has been demonstrated. Thus it is a useful way of applying Neural Networks or other classifiers in an intelligent manner. In passing, it should be noted that boosting techniques for Support Vector Machines and other statistical learning methods have also been developed [175, 135, 110].

The AdaBoost algorithm is as follows:

ADABOOST ALGORITHM
$t = 0$;
input: *A training population matrix* $\boldsymbol{X}$ *of size* $n \times d$ *with rows* $\boldsymbol{x}_i$, *labeled with* $c_i \in \{-$
input: *A maximum number of iterations* T
initialize $t = 0$;
initialize *the sample distribution:* $p_t(i) = 1/n, i = 1, 2, \ldots, n$;
while $t \leq T$
  $t = t + 1$;
  // *train a classifier* $c_t(\boldsymbol{x}_i)$
  on samples selected according to the discrete distribution $p_t(i)$;
  // *compute training error:*
  $e_t = [\sum_{i=1}^{n} p_t(i) \Upsilon(c_i \neq c_t(\boldsymbol{x}_i))] \,/\, [\sum_{i=1}^{n} p_t(i)]$;
  // *compute classifier weights:*
  $\gamma_t = \log\left(\frac{1-e_t}{e_t}\right)$;
  // *update sample distribution:*
  $p_{t+1}(i) = p_t(i) \exp[\gamma_t \Upsilon(c_i \neq c_t(\boldsymbol{x}_i))]$;
end while
return $c(\boldsymbol{x}) = \text{sgn}\left[\sum_{t=1}^{T} \gamma_t c_t(\boldsymbol{x})\right]$

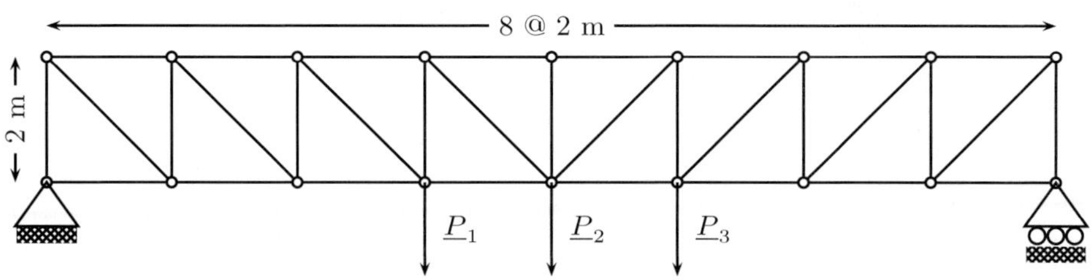

Fig. 6.3. Truss example solved with the AdaBoost algorithm

As an example consider the truss depicted in Figure 6.3. The structure is subject to three random loads, namely $\underline{P}_1, \underline{P}_2$ and $\underline{P}_3$. Also the Young moduli corresponding to the upper and lower chords and to the diagonal and vertical members, i.e. $\underline{E}_{\mathrm{uc}}$, $\underline{E}_{\mathrm{lc}}$, $\underline{E}_{\mathrm{diag}}$ and $\underline{E}_{\mathrm{vert}}$, respectively, are random. The statistical properties of these random variables are shown in Table 6.2.

Table 6.2. Statistical properties of the random variables of the example truss

Variable	Distribution	Mean	Standard deviation
$\underline{P}_1$	Gumbel	98.1 kN	9.81 kN
$\underline{P}_2$	Gumbel	98.1 kN	9.81 kN
$\underline{P}_3$	Gumbel	98.1 kN	9.81 kN
$\underline{E}_{\mathrm{uc}}$	Lognormal	200000 MPa	2000 MPa
$\underline{E}_{\mathrm{lc}}$	Lognormal	200000 MPa	2000 MPa
$\underline{E}_{\mathrm{diag}}$	Lognormal	200000 MPa	2000 MPa
$\underline{E}_{\mathrm{vert}}$	Lognormal	200000 MPa	2000 MPa

It is assumed that the truss fails if the displacement of the center node of the lower chord $\delta(\mathbf{X})$ is greater than 0.0173 m, so that the limit state function can be written as

$$g(\boldsymbol{x}) = 0.0173\,\mathrm{m} - \delta(\boldsymbol{x}) \,, \tag{6.5}$$

where $\boldsymbol{x} = [P_1, P_2, P_3, E_{\mathrm{uc}}, E_{\mathrm{lc}}, E_{\mathrm{diag}}, E_{\mathrm{vert}}]^{\mathrm{T}}$. The probability of failure was calculated by means of a simple Monte Carlo sampling using 300,000 samples, and a committee of five Multi-Layer Perceptrons with four neurons in the hidden layer equipped with sigmoid functions and one neuron in the output layer applying a purely linear mapping. The networks were trained with a data set consisting of 143 samples of an exploratory survey of the input variable space and 150 samples of a directional sampling oriented to the estimated design point. Thus each MLP was trained with the same data set of 293 points using the Levenberg-Marquardt back-propagation algorithm and the decision of the committee was calculated with the AdaBoost algorithm. With the Monte Carlo simulation a probability of failure of 1.0033×10^{-3} was obtained whereas the committee of MLPs yielded a probability of 1.0100×10^{-3} over the same set of 300,000 samples used for testing. This means that the committee, in classifying, failed to assess only two samples. It can be seen the error in the computation of the probability of failure by the committee of MLPs is almost negligible.

6.3.2 A Second Comparison of MLP and RBFN

In Chap. 4 it was shown that Radial Basis Function Networks are not convenient for solving structural reliability problems under the classification viewpoint and that to this end Multi-Layer perceptrons are more indicated. This is due to the different kind of mappings they perform. In this section, some further comparisons are made, with regard to the application of these network types to the functional approximation needed in structural reliability.

It was said in Chap. 4 that the conventional way of using RBFN is by means of a greedy selection of the radial basis function centers using statistical error criteria. Another way to choose the centers is by applying Vector Quantization or Expectation-Maximization algorithms. Since these are unsupervised learning techniques, their use for producing the small subset of RBF centers out of a large set of training samples implies a mixed unsupervised-supervised approach for training the network. For structural reliability analysis this could have an important consequence: the number of training samples actually processed by a finite-element code can be reduced. This is due to the fact that a large amount of random samples of the basic variables can be used to select a small number of centers. These centers are then processed by a finite-element code in order to obtain the value of the performance function for calculating the weights of the output connections in the supervised mode. This possibility is not directly available in MLP, with the consequence that there is a higher risk of computation redundancy in their application with entirely random samples.

This explains why in the experiments reported by the author in [79] the RBFN showed much better properties for a blind approximation of the performance function and, consequently, the probability of failure in complex structural systems. By a blind approximation we mean an estimation in which purely random samples are used and no information is assumed to exist on the critical orthants and the design point, which would facilitate the generation of a training sample set. Of course, when this information is available, any of the two kinds of networks used as functional approximators reports good results, because the training set can be generated with Importance Sampling [123] or with Directional Simulation [77]. However, gaining such a previous information requires an exploration with samples that in high-dimensional spaces is harassed by the curse of dimensionality. In order to avoid computation redundancy with MLP, they can also be trained after applying a data compression method whose result is a set of well separated samples.

Another point of comparison between MLP and RBFN is that of model complexity. According to the theory of statistical learning sketched in Chap. 2, the VC-dimension of a linear model like (4.43) is $m+1$ where m is the number of constituent functions, which equals the number of neurons in the hidden layer. Comparing this figure with the upper bound of the VC-dimension of an MLP (4.31), it seems that the RBFN model is much less complex. However, for a good approximation they normally require many neurons in the hidden layer as shown next, so that their complexity is in fact similar to that of MLP. Finally, the linearity of its model allows a simple calculation of SORM quantities, as shown in Chap. 1, whereas MLP require complicated calculations for the same purpose, as described at length by [16].

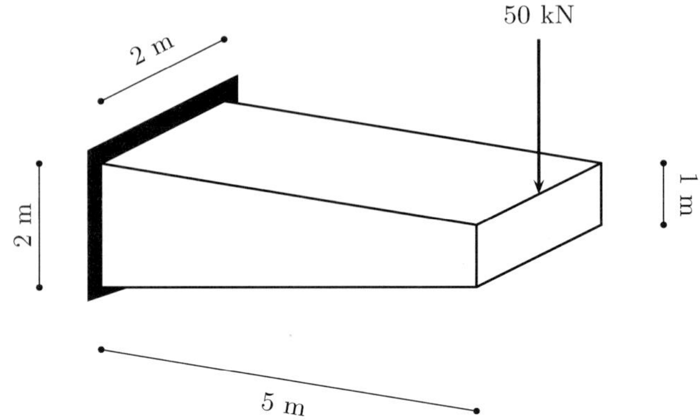

Fig. 6.4. A tapered beam for a stochastic finite element example

6.3.3 Example: Full Probabilistic Analysis with Stochastic Finite Elements

Applications of neural networks for a full probabilistic analysis using stochastic finite elements are scarce. The author has reported some analyses using Radial Basis Function Networks [82, 78]. In this section a comparison is made between MLP and RBFN over a simple but illustrative example.

Let us consider the tapered cantilever beam shown in Fig. 6.4. It has an applied load $p = 50$ kN and a linearly varying cross section height. The modulus of elasticity E is a random process with mean $\mu = 1 \times 10^7$ and autocorrelation function

$$B(r, s) = \sigma^2 \exp[-\frac{1}{c}|r - s|] \ , \tag{6.6}$$

in which $\sigma = 2 \times 10^6 \text{kN/m}^2$ and $c = 3$. Notice that the spread of the process, as measured by $\sigma/\mu = 0.2$, is high. The beam was discretized into 16 finite elements. In order to reduce the number of dimensions of the stochastic problem the random field was represented with the Karhunen-Loève expansion summarized in Chap. 3. For the solution of the Karhunen-Loève problem use was made of the Nyström method [6], which is based on a Gaussian quadrature approximation of the integral equation 3.37. The approximation was performed

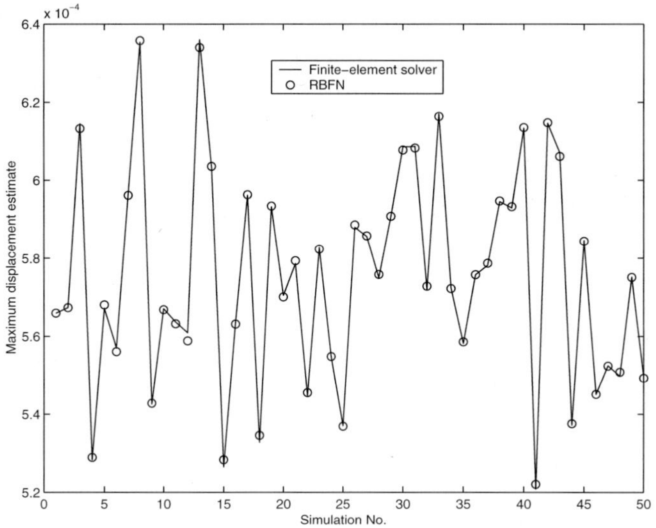

Fig. 6.5. Comparison of RBFN estimates of the tapered beam displacement with the actual values

with nine Gauss-Legendre quadrature points, which allow the determination of an equal number of eigenvalues that are sufficient for the present problem[1]. For each Monte Carlo simulation a Karhunen-Loève realization of the $E-$random field comprising 256 points was performed and the value of the E of each element was obtained as the average of 16 points contained in it, according to the local average technique [173].

Two thousand Monte Carlo simulations were performed for comparing the performance of MLP and RBFN. These were trained with the same subset of 150 samples and the rest were used for testing. The only output considered was the maximum displacement of the beam. Notice that since the relationship of the displacement and the modulus of elasticity is nonlinear, and this latter is defined by a random vector of nine components, the mapping is highly nonlinear despite the simplicity of the problem from the mechanical viewpoint.

[1]The main advantage of the Nyström method is that the number of eigenvalues, which need not be large in most stochastic finite element application, can be made equal to the number of quadratures. This implies a reduction of the size of the matrices involved in comparison to other techniques, which require the solution of large matrix problems for the same number of eigenvalues. Its main disadvantage is its difficulty to deal with complex physical domains.

The architecture of the MLP is as follows: nine neurons in the input layer corresponding to the Karhunen-Loève values u_i in (3.40); twelve neurons in the hidden layer with hyperbolic tangent activation function; and one neuron in the output layer with a purely linear function yielding the estimate of the maximum displacement. For the RBFN, the type used was that of the algebraic type (see Sect. 4.5), meaning that the hidden layer comprehends 150 neurons with Gaussian activation function. It was adopted because the least-square type with few neurons yielded poor results[2]. The training times required by the RBFN was less than a second and for the MLP about 30 sec. Such a fast training makes it possible to calculate several alternatives with a little increase of the computational effort in order to select the best architecture in each application.

Figures 6.5 and 6.6 show a sample-by-sample comparison between the NN estimates and the exact, i.e. finite-element given values for RBFN and MLP, respectively. An excellent accuracy in both cases is evident, being higher for the MLP. The small differences in the RBFN case have an effect in the estimates of the probability distribution function appearing in Fig. 6.7. The corresponding result for MLP appears under Fig. 6.8. Since for reliability analysis purposes the largest values are of higher importance, a further comparison is made of the complementary probability at the upper tail in Figs.

[2]This illustrates that the complexity of the RBFN can indeed be high as said above.

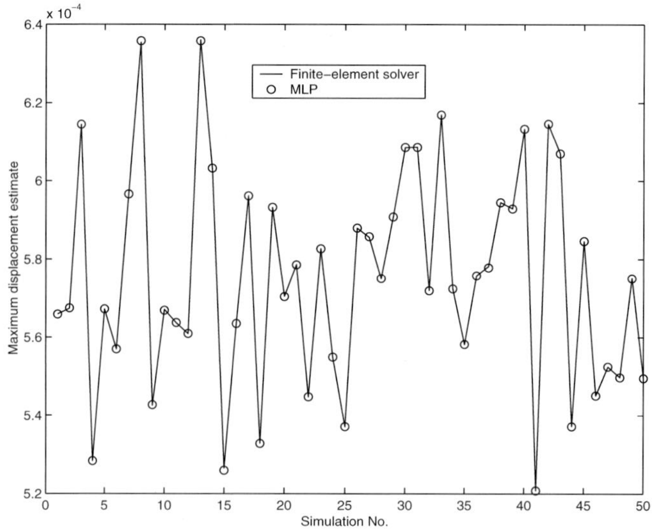

Fig. 6.6. Comparison of MLP estimates of the tapered beam displacement with the actual values

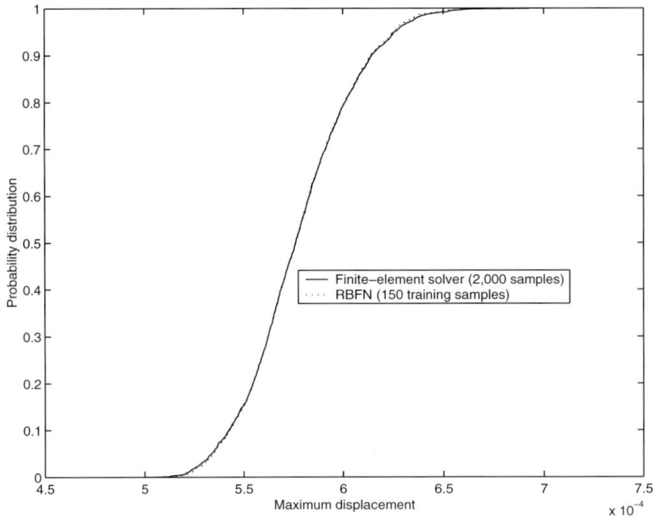

Fig. 6.7. RBFN probability distribution estimates

6.9 and 6.10. On the basis of these figures it can be concluded that the MLP are superior than the RBFN in this case.

6.4 Support Vector Regression

The approximation of real functions using Support Vector Machines is based on the treatment of non-separable classes, which in turn generalizes the perfect separable case dealt with in Chap. 5. In this section we will first summarize these issues and evaluate the accuracy of Support Vector Regression (SVR) with a practical example.

6.4.1 Support Vector Approach to Non-Separable Classes

The extension of the support vector classification methodology to non-separable classes was developed by [30] by introducing the so-called *slack variables* $\xi_i \geq 0, i = 1, 2, \ldots, n$ associated to each of the n training samples (Fig. 6.11) allowing the constraints (5.4) to take the relaxed form

$$c_i(\langle \boldsymbol{w}, \boldsymbol{x}_i \rangle - b) \geq 1 - \xi_i \ . \tag{6.7}$$

With this change the optimization problem becomes

$$\text{minimize } \Omega(\boldsymbol{w}) = \frac{||\boldsymbol{w}||^2}{2} + C\sum_{i=1}^{n} \xi_i$$

$$\text{subject to: } c_i(\langle \boldsymbol{w}, \boldsymbol{x} \rangle - b) \geq 1 - \xi_i, i = 1, 2, \ldots, n. \qquad \xi_i \geq 0 \ ,$$

meaning that a solution $(\boldsymbol{w}, b)$ is sought for a constant $C > 0$ such that the margin is maximal and the number of classification errors is minimal. In fact, the term $\sum \xi_i$ is a relaxed form of the actual sum of misclassifications given by $\sum \Upsilon(\xi_i > 0)$ [27]. The function that realizes the solution of this optimization problem in the feature space is called *soft-margin hyperplane*. After proceeding in the same manner as in Chap. 5, the same classifier (5.16) is obtained, with the new restriction $0 \leq \alpha_i \leq C$. Thus the constant C decides the trade-off between the complexity and the percent of non-separable training samples. Therefore, it is assigned by the analyst. Finally, the threshold b can be computed with (5.21) because the slack variables for samples with $\alpha_i < C$ are null. See [175] for the details.

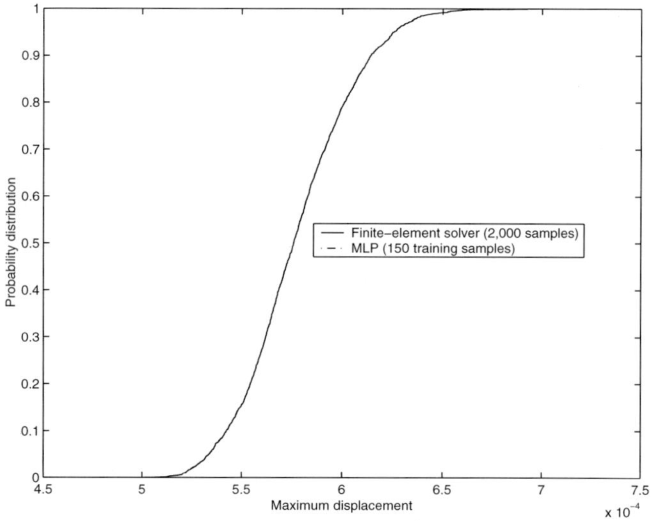

Fig. 6.8. MLP probability distribution estimates

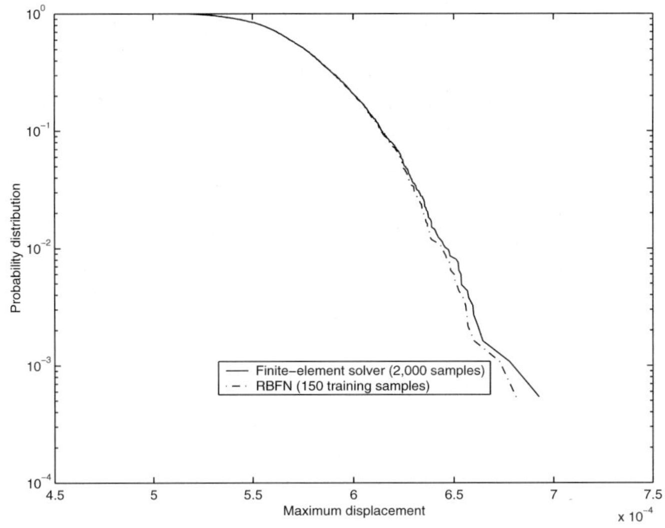

Fig. 6.9. Upper tail of RBFN probability distribution estimates

6.4.2 Extension to Function Approximation

Let $(\boldsymbol{x}_i, y_i), i = 1, 2, \ldots, n, \boldsymbol{x}_i \in \mathbb{R}^d, y_i \in \mathbb{R}$ denote the training samples pairs for a function approximation problem. The function to be fitted has the linear form

$$f(\boldsymbol{x}) = \langle \boldsymbol{w}, \boldsymbol{x} \rangle - b \, , \tag{6.8}$$

which after the nonlinear projection to the feature space using the implicit mapping of kernels will hold in that space as well. The extension of the support vector methodology for classification to this situation is based on the soft-margin concept just explained. For regression analysis, the soft-margin is transformed into the so-called $\epsilon-$*insensitive loss function* defined as (see Fig. 6.12)

$$|y - f(\boldsymbol{x})|_\epsilon = \max(0, |y - f(\boldsymbol{x})| - \epsilon) \, , \tag{6.9}$$

where ϵ is a parameter controlling the precision of the regression. The parameter generates an $\epsilon-$tube such that if it is too narrow many points will be outside it. In order to remedy this, slack variables ξ, ξ^* similar to those used in non-separable classification are introduced and the optimization problem is formulated as

$$\text{minimize } \Omega(\boldsymbol{w}, \boldsymbol{\xi}, \boldsymbol{\xi}^*) = \frac{||\boldsymbol{w}||^2}{2} + C \sum_{i=1}^{n} (\xi_i + \xi_i^*)$$

$$\text{subject to: } y_i - \langle \boldsymbol{w}, \boldsymbol{x}_i \rangle + b \leq \epsilon + \xi_i, \quad i = 1, 2, \ldots, n \ ,$$

$$\langle \boldsymbol{w}, \boldsymbol{x}_i \rangle - b - y_i \leq \epsilon + \xi_i^* \ ,$$

$$\xi_i, \xi_i^* \geq 0 \ ,$$

where the role of constant C is the same as above, namely to control the trade-off between the complexity and the portion of errors that are tolerated to within an accuracy ϵ. The Lagrange function for this latter problem is

$$L(\boldsymbol{w}, b, , \boldsymbol{\xi}, \boldsymbol{\xi}^*, \boldsymbol{\alpha}, \boldsymbol{\alpha}^*, \boldsymbol{\eta}, \boldsymbol{\eta}^*) = \frac{1}{2} \|\boldsymbol{w}\|^2 + C \sum_{i=1}^{n} (\xi_i + \xi_i^*)$$

$$- \sum_{i=1}^{n} \alpha_i (\epsilon + \xi_i - y_i + \langle \boldsymbol{w}, \boldsymbol{x}_i \rangle - b) \tag{6.10}$$

$$- \sum_{i=1}^{n} \alpha_i^* (\epsilon + \xi_i^* + y_i - \langle \boldsymbol{w}, \boldsymbol{x}_i \rangle + b) - \sum_{i=1}^{n} (\eta_i \xi_i + \eta_i^* \xi_i^*) \ ,$$

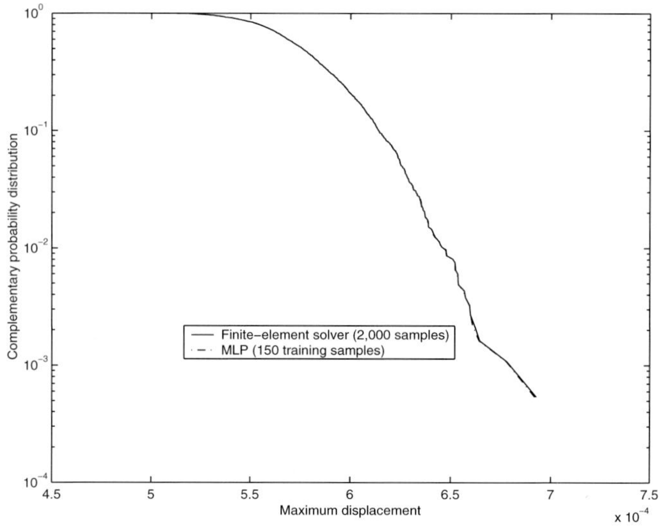

Fig. 6.10. Upper tail of MLP probability distribution estimates

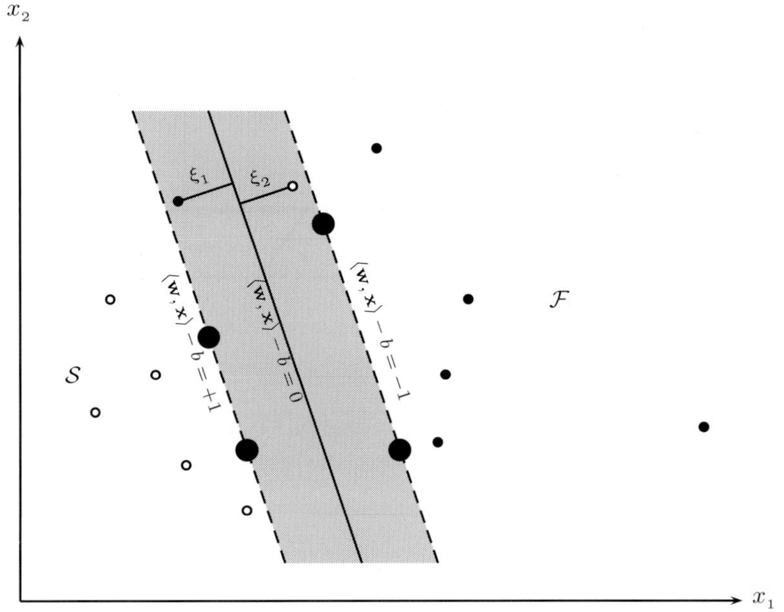

Fig. 6.11. Slack variables and soft margin for a non-separable classification problem. Notice the presence of samples of each class on the wrong side of the hyperplane

with multipliers $\alpha_i, \alpha_i^*, \eta_i, \eta_i^* \geq 0$. Similarly to the classification case the solution sought is a saddlepoint, in which the maximization is performed with respect to the set $(\boldsymbol{\alpha}, \boldsymbol{\alpha}^*, \boldsymbol{\eta}, \boldsymbol{\eta}^*)$ and the minimization with respect to $(\boldsymbol{w}, b, \boldsymbol{\xi}, \boldsymbol{\xi}^*)$. Upon setting to zero the partial derivatives of the Lagrangian with respect to the latter set yields

$$\begin{aligned} \boldsymbol{w} &= \sum_{i=1}^{n} (\alpha_i - \alpha_i^*) \boldsymbol{x}_i \,, \\ &\sum_{i=1}^{n} (\alpha_i - \alpha_i^*) = 0 \,, \\ &C - \alpha_i - \eta_i = 0 \,, \\ &C - \alpha_i^* - \eta_i^* = 0 \,. \end{aligned} \tag{6.11}$$

Substituting the first of the above equations into (6.8) gives the solution

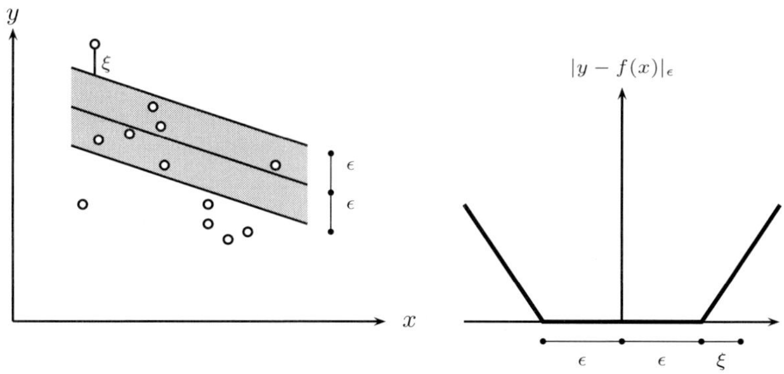

Fig. 6.12. ϵ−insensitive loss function used in Support Vector Regression

$$f(\boldsymbol{x}) = \sum_{i=1}^{n} (\alpha_i - \alpha_i^*) \langle \boldsymbol{x}, \boldsymbol{x}_i \rangle - b \,, \tag{6.12}$$

where the threshold b is computed with $\xi_i = 0$ in the condition $y_i - \langle \boldsymbol{w}, \boldsymbol{x}_i \rangle + b \leq \epsilon + \xi_i$, which becomes an equality for those samples having $0 \leq \alpha_i \leq C$, as in the non-separable classification case. A similar reasoning applies to the constraint $\langle \boldsymbol{w}, \boldsymbol{x}_i \rangle - b - y_i \leq \epsilon + \xi_i^*$.

The extension of (6.12) to nonlinear functions is straightforward using kernels:

$$f(\boldsymbol{x}) = \sum_{i=1}^{n} (\alpha_i - \alpha_i^*) K(\boldsymbol{x}, \boldsymbol{x}_i) - b \,. \tag{6.13}$$

Notice that as in the pattern recognition problem, the approximation of a function $f(\boldsymbol{x}), \boldsymbol{x} \in \mathbb{R}^d$ does not depend on the dimensionality degree d.

6.4.3 Example: Random Eigenvalues of a Frame

The accuracy of this regression approach for structural reliability applications is evaluated in this section with a simple example consisting in estimating the probability distribution of the first eigenfrequency of the frame appearing in Fig. 6.13. The modulus of elasticity is $E = 2 \times 10^7 \text{kN/m}^2$, the moment of inertia is $I = 1/96 \text{m}^4$ and the length $l = 3$ m.

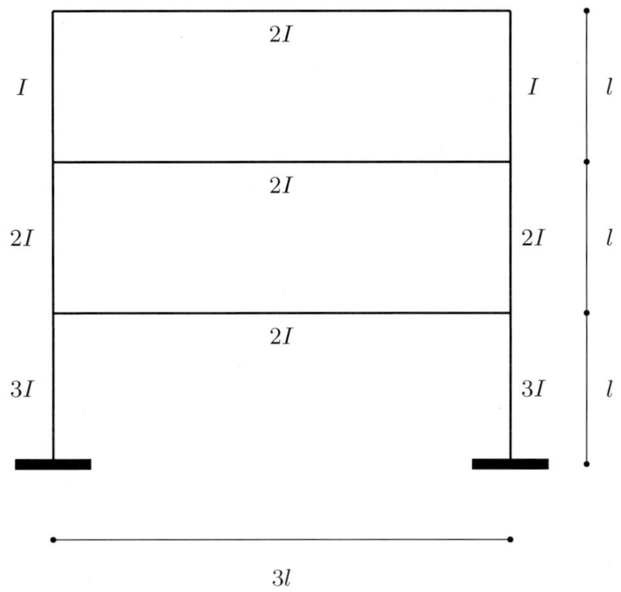

Fig. 6.13. Frame example for Support Vector Regression analysis

Two analyses were conducted. In the first the only random variable is a mass assigned to all the three floors simultaneously. It is defined as a Normal variable with mean 35 t and a coefficient of variation equal to 0.15. In the second the three masses were independent Normal variables with the same statistics. In both cases the Support Vector Regressor was trained with 150 random samples with the values $C = 1,000$ and $\epsilon = 0.0001$. It was tested with 850 samples, making a total of 1,000 samples. A comparison of the actual and estimated outputs for all the samples in the first case is shown in Fig. 6.14, in which the first 150 samples correspond to the training set. It can be observed that in the left part the coincidence is almost perfect and that there are some minor departures from the actual values in the test portion of the picture. In Fig. 6.15 there is a log-plot of the probability and complementary probability distribution functions of the eigenfrequency. For the graph emphasizing the lower tail there are no differences between the actual function and its estimate, while for that of the upper tail there is a departure that is remedied towards its end.

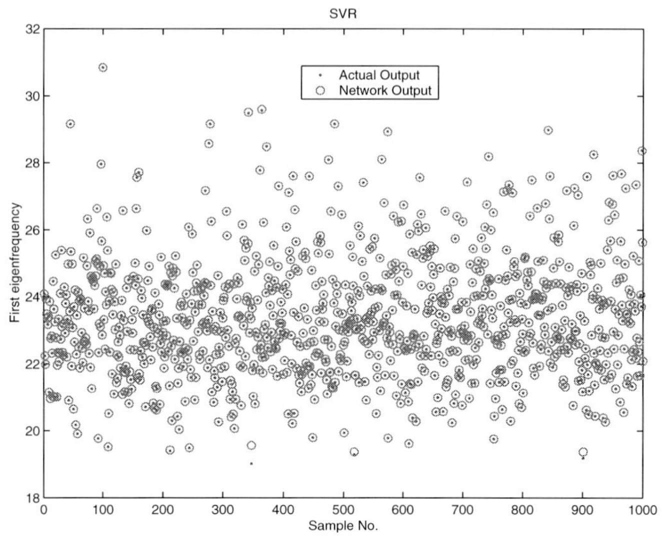

Fig. 6.14. Comparison of SVR frequency estimates with actual values for the problem with one random mass

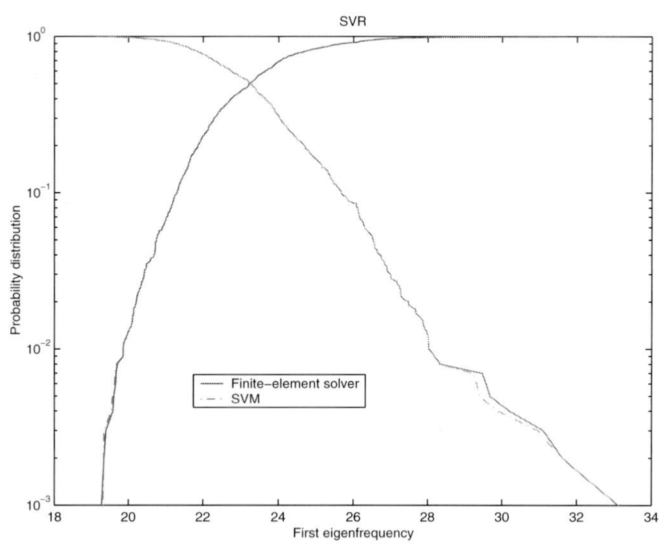

Fig. 6.15. SVR frequency distribution function estimates at lower and upper tails

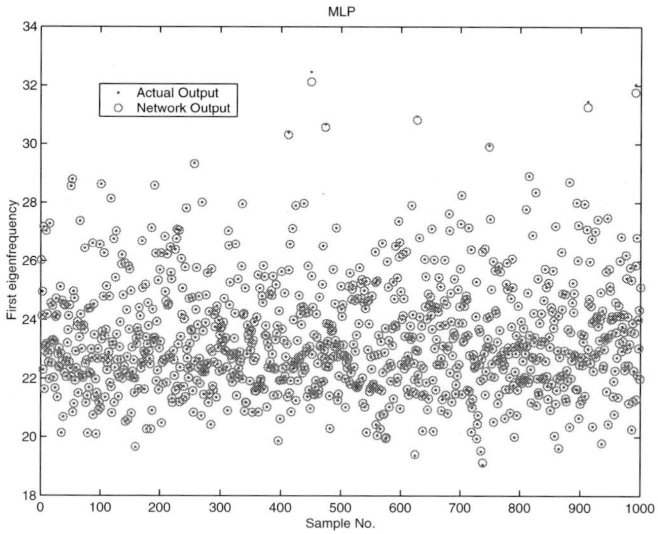

Fig. 6.16. Comparison of SVR frequency estimates with actual values for the problem with one random mass

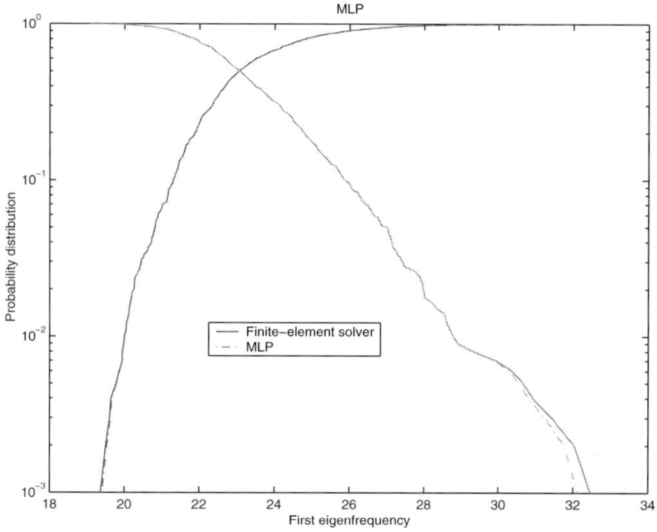

Fig. 6.17. MLP frequency distribution function estimates at lower and upper tails

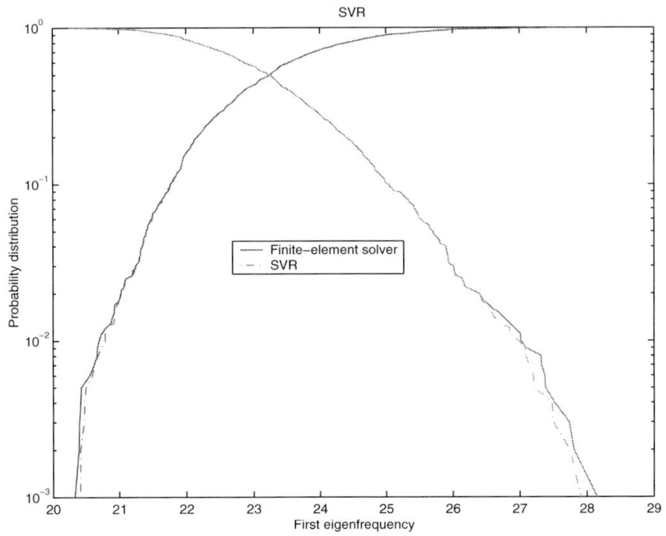

Fig. 6.18. SVR frequency distribution function estimates at lower and upper tails. Frame with three random masses

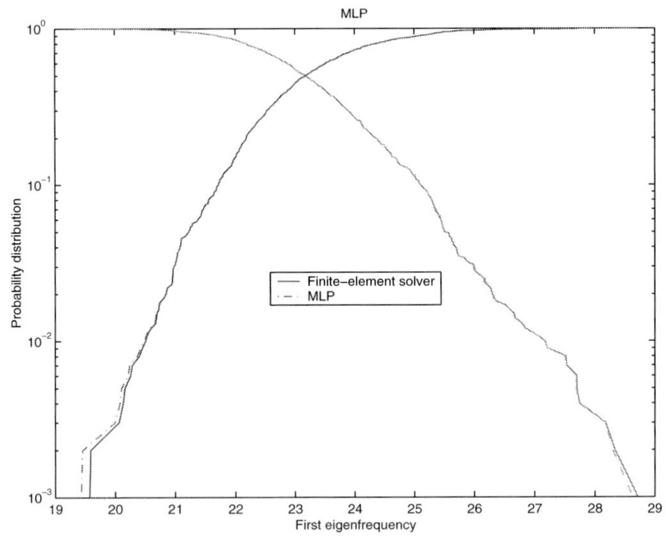

Fig. 6.19. MLP frequency distribution function estimates at lower and upper tails. Frame with three random masses

For comparison a similar analysis was made with a MLP having six neurons in the hidden layer with hyperbolic tangent activation and a single neuron in the output layer with a purely linear function. The results appear in Figs. 6.16 and 6.17. It is evident that the SVM exhibits a superior behavior in this example. The results for the second case involving three independent random masses are shown in Figs. 6.18 and 6.19. In this case, the MLP shows a somewhat superior behavior. What is entirely evident from these examples and previous ones is that both techniques are very adequate for building a surrogate of the finite-element solver.

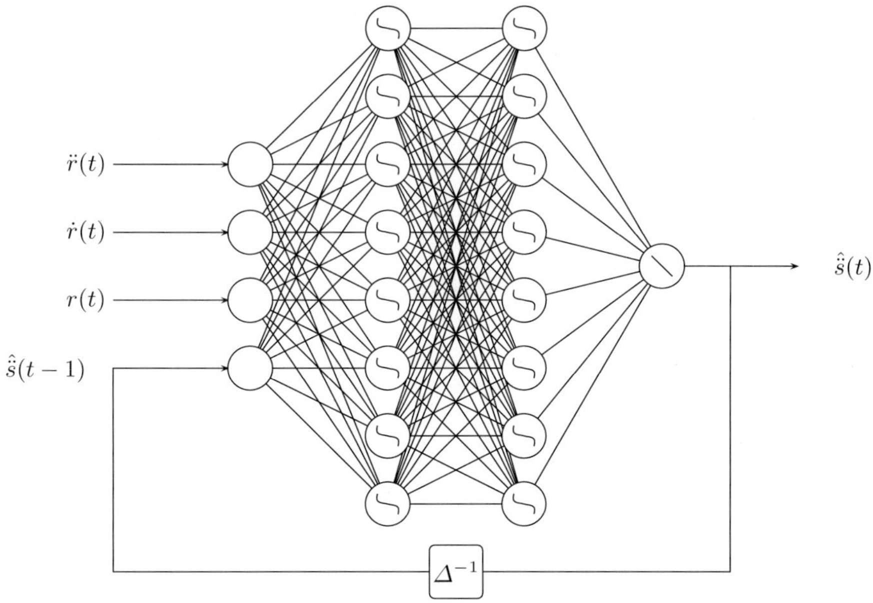

Fig. 6.20. Time-dependent MLP used for signal mapping (Adapted from [84])

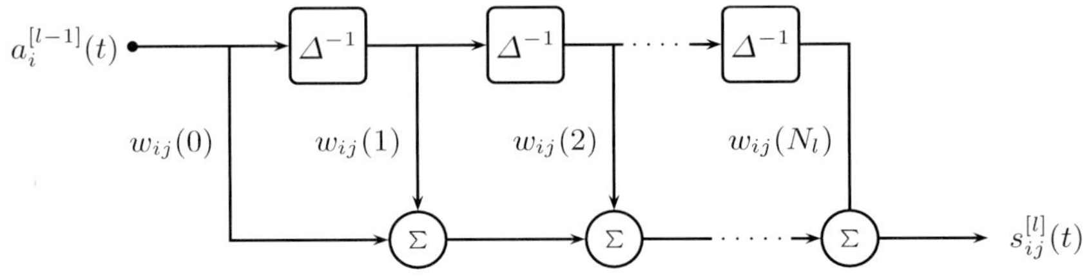

Fig. 6.21. Scheme of application of the lag operator (Adapted from [84])

6.5 Time-Dependent MLP for Random Vibrations

The estimation of dynamic input/output mappings occurring in structural and soil dynamics is a more difficult problem than those static ones faced in the preceding, due to the influence of velocity and acceleration terms. These induce a dependence of the output not only on the current values of the input but also on the previous values of the output itself. A naive use of neural networks such as MLP and RBFN in this case would lead to large errors. In this section neural networks intended to deal with these situations and used by the author in [84] are summarized together with the soil dynamic applications reported therein. This type of network will be called time-dependent MLP (TD-MLP).

Conventional MLP described in the preceding have successfully been used in several applications requiring time-independent mappings of input onto output variables. As a time-dependent application, they have also proved to be useful in system identification, especially in forecasting future values of a time series being the output of a dynamical system, which is an important task in automatic control. Roughly speaking, this is a situation in which the interest lies in estimating the system output at time $(k + n)$ on the basis of the current input as well as the actually recorded output values at time k. After these output values are computed they are used to excite the control mechanisms but are not fed back to the network; instead, the new recorded values are used. This way of using the recording information is known, in System Identification and Control Theory, as *series* or *open-loop training.*

Although, in stochastic mechanics applications such as those considered in this book, the purpose is to use the network for simulating signals not ever recorded and which will never be so, because of the virtuality of their realization. Therefore, the estimated output values must be fed back through the network.

For performing a signal mapping in this way conventional MLP are not adequate because there is an accumulation of errors, as has been shown [84]. In order to overcome this difficulty, some approaches have been proposed, such as recurrent feedforward networks [107] and finite impulse response (FIR) networks [182]. These later are described next as they are useful for a complete nonlinear signal mapping in a random vibration problem.

These TD-MLP are inspired in the fact that certain operations performed by the brain recall the performance of linear, finite impulse response (FIR) filters. TD-MLP allow the output to be backfed through the first layer together with some delayed inputs, which is the aim for an off-line input-output signal mapping (see Fig. 6.20). In the figure, Δ^{-1} stands for a single-step lag operator. In essence, the main difference of this network type with respect to classical MLP is that the input to a neuron is obtained as a discrete convolution, in a similar manner as in standard FIR theory (see e.g. [127]) :

$$\begin{aligned} s_j^{[l]}(k) &= \textstyle\sum_{i=1}^{I_{l-1}} \sum_{n=0}^{N_l} w_{ij}^{[l]}(k)\ a_i^{[l-1]}(k-n) \\ &= \textstyle\sum_{i=1}^{I_{l-1}} \boldsymbol{w}_{ij}^{[l]} \cdot \boldsymbol{a}_i^{[l-1]}(k)\ . \end{aligned} \tag{6.14}$$

Here

$$\boldsymbol{w}_{ij}^{[l]} = \left[w_{ij}(0), w_{ij}(1), \ldots, w_{ij}(N_l)\right] \tag{6.15}$$

is the vector of weights associated to the $l-$th layer and

$$\boldsymbol{a}_i^{[l]}(k) = \left[a_i^{[l]}(k), a_i^{[l]}(k-1), \ldots, a_i^{[l]}(k-N_{l+1})\right] \tag{6.16}$$

is the vector of delayed inputs. The indices k and n stand for time. This equation indicates that the input to the neuron j at time k is a linear combination of the impulses sent by the neurons of the back layer at the current time and some previous instants, as illustrated by Fig. 6.21. Therefore, this neural network has spatially as well as temporally connected neurons. This feature implies an enhancement of its memory capabilities.

The way of training is in all respects similar to that of the time independent MLP summarized in Chap. 4, with the exception that the resulting equations are more involved because of the time-dependence of the connections. The end training equations are

$$\boldsymbol{\Delta w}_{ij}^{[l]}(k) = -\eta \delta_j^{[l]}(k) \boldsymbol{a}_i^{[l-1](k)}\ , \tag{6.17}$$

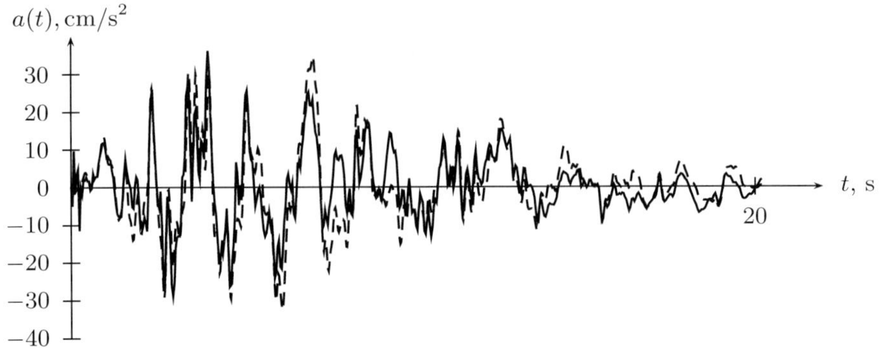

Fig. 6.22. Time-independent MLP trained with acceleration signals. The solid line shows an actual free-surface signal. The dashed line is its MLP estimate obtained by presenting the network the corresponding bedrock acceleration signal. The training was made with a different pair of bedrock and free-surface records

$$\delta_j^{[l]} = \begin{cases} -2e_j f'(s_j^{[L]}), & l = L \\ f'(s_j^{[l]}) \sum_{i=1}^{I_{l+1}} \boldsymbol{\delta}_i^{[l+1]}(k) \cdot \boldsymbol{w}_{ji}^{[l+1]}, & 1 \le l \le L-1 \,. \end{cases} \quad (6.18)$$

Here $\boldsymbol{\delta}_i^{[l]}(k)$ is the vector

$$\boldsymbol{\delta}_i^{[l]}(k) = \left[\boldsymbol{\delta}_i^{[l]}(k), \boldsymbol{\delta}_i^{[l]}(k+1), \ldots, \boldsymbol{\delta}_i^{[l]}(k+N_{l+1})\right] \quad (6.19)$$

and $f(\cdot)$ is the activation function of the corresponding neuron. The main difference of these equations with respect to those of the time independent MLP lies in that some quantities become vectors as shown. In this formulation it has been assumed that the number of delays in each layer l, N_l, is the same for all of its neurons. See [182] for the details.

The neural networks just described were used for estimating the mapping of an input onto the output signals of a nonlinear system, which in this case is a column of soil strata [84]. The synthetic bedrock records were generated after a Kanai-Tajimi model. The corresponding one-sided spectral density function is

$$G(\omega) = \frac{\omega_g^4 + 4\nu_g^2\omega_g^2\omega^2}{(\omega_g^2 - \omega^2)^2 + 4\nu_g^2\omega_g^2\omega^2} G_0 \,. \quad (6.20)$$

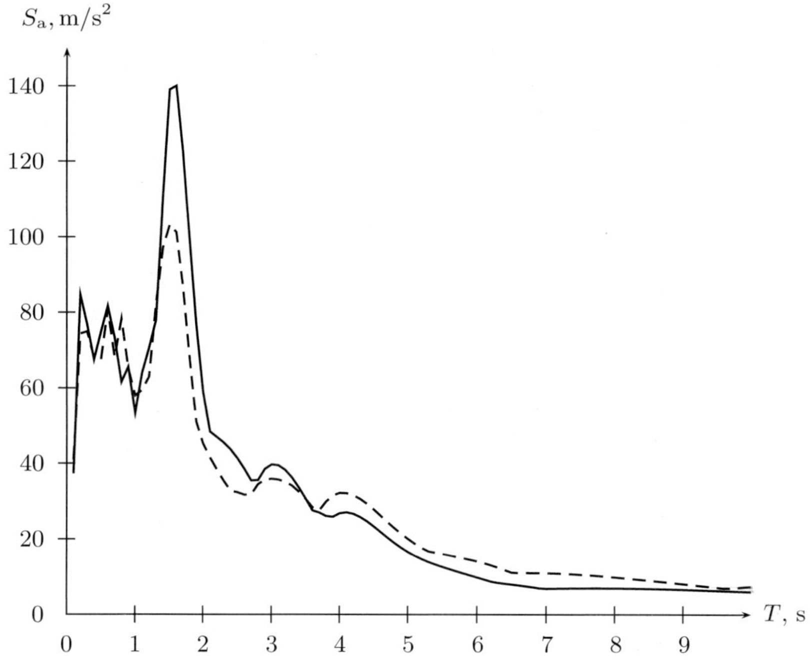

Fig. 6.23. Acceleration spectra of the signals in Figure 6.22

A realistic nonstationarity of the bedrock history may be obtained by multiplying the stationary signal generated with this spectral density by a modulating function such as this [155]:

$$\xi(t) = \frac{1}{c}\left(\mathrm{e}^{-at} - \mathrm{e}^{-bt}\right) \tag{6.21}$$

where a and b are parameters and

$$c = \max\left(\mathrm{e}^{-at} - \mathrm{e}^{-bt}\right) = \left[\left(\frac{a}{b}\right)^{a/(b-a)} - \left(\frac{a}{b}\right)^{b/(b-a)}\right]. \tag{6.22}$$

The values used for the parameters were $\omega_{\mathrm{g}} = 16$ rad/s, $\nu_{\mathrm{g}} = 0.64, G_0 = 0.4$ g$^2 \cdot$ s, $a = 0.25, b = 0.5$. The algorithm employed for simulation was the well known spectral representation method [154]. The peak acceleration of the bedrock signals was not restricted to a specified value in order to test the accuracy of the network under the influence of the randomness of this parameter, in the limits given by the stochastic spectral model. A soil dynamics code was used to obtained the free-surface records of an unidimensional profile consisting of five layers of sand, gravel and clay. The network used has two hidden layers of neurons activated with the hyperbolic tangent function. The first layers has with 8 neurons with 6 delays each, and the second 8 neurons with 4 delays. The training was performed with a single input/output pair of synthetic signals. One pair of bedrock and free-surface signals were used in the training phase. Figure 6.22 compares the actual and estimated free-surface signals corresponding to a bedrock signal different from that used in the training. It can be seen that the agreement is fairly good. This agreement suggests the possibility of employing the network as a surrogate of the soil-dynamics solver for estimating the $\mu + \sigma$ (mean plus one standard deviation) free-surface spectrum which is highly useful in earthquake engineering practice. To this end, fifty synthetic bedrock records were simulated and used as input to the soil-dynamics program, on the one hand, and to the trained network, on the other. The results for the $\mu+\sigma$ spectrum are compared in Fig. 6.23. It is readily seen that the assessment given by the network is excellent from a practical point of view.

7

Classification Approaches to Reliability Indexation

7.1 Introduction

In structural safety the computation of the probability of failure obviously requires the knowledge of the joint density function of the basic variables $\underline{\boldsymbol{x}}$. In some cases, however, only the marginal distributions and, perhaps, the correlation matrix are known, with the consequence that approximate models for the joint density function should be applied [36]. A more frequent situation is that even the marginal densities remain unknown, due to the insufficiency of experimental evidence [115] and hence only low order marginal and joint moments can be confidently estimated. In such situation the estimation of the reliability as

$$R = \int_{\mathcal{S}} f_{\underline{\boldsymbol{x}}}(\boldsymbol{x})\mathrm{d}\boldsymbol{x} \tag{7.1}$$

can only be performed on some strong assumptions, such as the linearity of the limit state function and the Gaussian behavior of the basic variables, because this model is entirely defined in terms of second order moment information. Since this is evidently a disputable decision, some authors propose to measure the reliability in terms of an index, which affords a meaningful value for design, as it measures the extent of the safe domain in some sense. It is specially useful when the statistical information on the basic variables is reduced, normally consisting of second-order marginal and cross moments of the basic variables. Since this is a common situation in structural reliability practice some authors prefer to formulate reliability approaches to practical problems (code calibration, load combination, reliability updating, etc.) in terms of it.

There are several proposals for the construction of meaningful reliability indices, some of which have entered the standard practice of structural engineering. These are the simple, geometric, generalized and empirical SORM indices. They are also known under the names of Cornell, Hasofer-Lind, Ditlevsen and Zhao-Ono, respectively, according to their developers [29, 68, 38, 187]. As a general rule, it can be said that the most widely applied is the Geometric index, despite it fails to feature some ideal properties a reliability index should have. This has promoted the development of the last two proposals of the above list.

In this chapter it is shown that the classification paradigm offers the possibility of building other alternatives which fulfill the ideal properties and that are simple to calculate. These are as follows:

- An index that can be calculated with the Support Vector Machines in the input $\boldsymbol{u}-$space. It is defined as the distance from the origin to a secant hyperplane, instead to the tangent one calculated with FORM techniques, which is the basis of the geometric index.
- An index construed with Support Vector Machines in the feature $\boldsymbol{\zeta}-$space. It is also based on a hyperplane, but it is not measured as a distance to it but on the basis of the volume it generates.

The generalized index was proposed as a generalization of the simple reliability index [40]. Instead, the above proposals can be considered as extensions of the geometric index, which is the most widely used and has become a standard in structural reliability practice. This is because the proposed indices are geometrical measures derived from hyperplane constructions, similarly to Hasofer-Lind's.

The chapter begins with a general discussion on the indices mentioned above from the point of view of the ideal properties. Then the hyperplane formulations in structural reliability exposed in the previous chapters are summarized and compared in order to show the relevance of such geometrical construction and the similarities of their calculation methods. Finally, the two classification-based reliability indices are explained in detail and their advantages are highlighted.

7.2 A Discussion on Reliability Indices

The main indices that have been proposed in the international literature are the following:

- *Simple reliability index* [29]. It is calculated in terms of the low order moments for the specific case of a linear limit-state function:

$$g(\boldsymbol{u}) = a_0 + \boldsymbol{a}^{\mathrm{T}}\boldsymbol{u} \;, \tag{7.2}$$

where a_0 and the vector $\boldsymbol{a}$ are constants. The index is

$$\beta_{\mathrm{C}} = \frac{\mu_{\underline{g}}}{\sigma_{\underline{g}}} \;, \tag{7.3}$$

where $\mu_{\underline{g}}$ and $\sigma_{\underline{g}}$ are respectively the mean and standard deviation of the random variable $\underline{g}(\underline{\boldsymbol{u}})$. In geometrical terms, (7.2) corresponds to a hyperplane. It is easy to show that this is equivalent to

$$\beta_{\mathrm{C}} = \frac{a_0 + \boldsymbol{a}^{\mathrm{T}}\boldsymbol{\mu}}{\sqrt{\boldsymbol{a}^{\mathrm{T}}\boldsymbol{C}\boldsymbol{a}}} , \tag{7.4}$$

in which $\boldsymbol{\mu}$ and $\boldsymbol{C}$ are respectively the mean vector and the covariance matrix of the basic variables.

- *Geometric index* [68]. This proposal is aimed to overcoming the limitations of the above index to deal with nonlinear limit state functions. To this purpose it is necessary to work in a standard space. As said above, the geometric index is defined as the minimum distance to the transformed function $g(\boldsymbol{u}) = 0$:

$$\beta_{\mathrm{HL}} = \min_{\{\boldsymbol{u}|g(\boldsymbol{u})=0\}} \left(\boldsymbol{u}^{\mathrm{T}}\boldsymbol{u}\right)^{\frac{1}{2}} . \tag{7.5}$$

In the original input space it becomes

$$\beta_{\mathrm{HL}} = \min_{\{\boldsymbol{x}|g(\boldsymbol{x})=0\}} \left((\underline{\boldsymbol{x}} - \boldsymbol{\mu})^{\mathrm{T}}\boldsymbol{C}^{-1}(\underline{\boldsymbol{x}} - \boldsymbol{\mu})\right)^{\frac{1}{2}} . \tag{7.6}$$

- *Generalized index* [38]. This index is defined as

$$\beta_{\mathrm{D}} = \Phi^{-1}\left[\int_{\mathcal{S}} \phi(\boldsymbol{u})\mathrm{d}\boldsymbol{u}\right] , \tag{7.7}$$

where $\Phi(\cdot)$ is the standard Gaussian distribution and $\phi(\boldsymbol{u}) = \phi(u_1)\phi(u_2)\ldots\phi(u_d)$, $\phi(\cdot)$ being the standard Normal density and d the number of basic variables.

- *Empirical SORM reliability index* [188]. This proposal is based on empirical analyses carried out on (1.28). The result reads

$$\beta_{\mathrm{ZO}} = \begin{cases} -\Phi^{-1}\left[\phi(-\beta_{\mathrm{HL}}\left(1 + \frac{\phi(\beta_{\mathrm{HL}})}{R\Phi(-\beta_{\mathrm{HL}})}\right)^{q}\right], & K \geq 0 \\ \left(1 + \frac{2.5K}{2d-5R+25(23-5\beta_{\mathrm{HL}})/R^2}\right)\beta_{\mathrm{HL}} + 0.5K\left(1 + \frac{K}{40}\right), & K < 0 \end{cases} \tag{7.8}$$

where

$$q = -\frac{d-1}{2}\left(1 + \frac{2K}{10(1 + 2\beta_{\mathrm{HL}})}\right) . \tag{7.9}$$

These indices are now examined with respect to some ideal properties that a reliability index should exhibit, as follows:

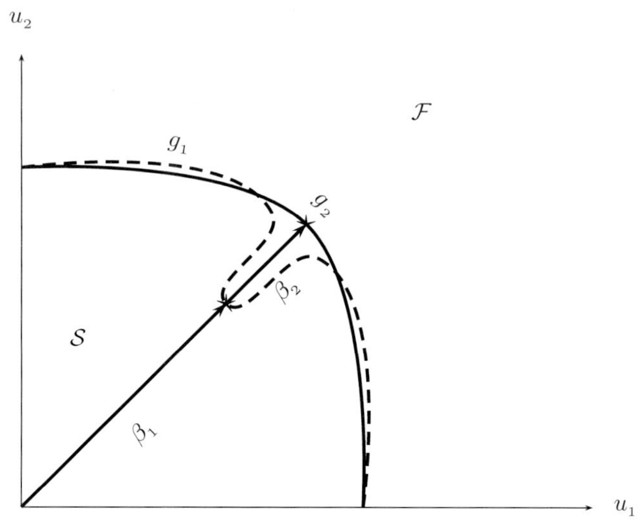

Fig. 7.1. Locality of the geometric reliability index. An admissible reliability index for function g_1 should be somewhere in between the two design points

- *Formulation invariance.* An admissible reliability index should not be dependent on the formulation of the limit-state function. In other words, if two limit-state functions coincide over an equal set of points, their associated reliability indices should be the same [40]. This is the case, for instance, of functions $g(\boldsymbol{u})$ and $g^3(\boldsymbol{u})$ that cross the $\boldsymbol{u}-$plane across the same contour. This property is satisfied by the geometric reliability index. The simple reliability index only satisfies it in the case of hyperplane limit states.
- *Globality.* The index should ideally be a global measure of the reliability associated to the entire safe domain and not be tied to a local detail of the limit-state function. The geometric index fails to satisfy this requirement, as it is tied to a single point (Fig. 7.1).
- *Ordering property.* The geometric reliability index also fails to exhibit this desirable property, which fostered the development of the generalized index [38]. Despite this property is a particular case of the previous one, it occupies a special place because it concerns to the reliability indices of safe domains forming a nested sequence $\mathcal{S}_1 \subset \mathcal{S}_2 \subset \mathcal{S}_3 \subset \ldots$ for which

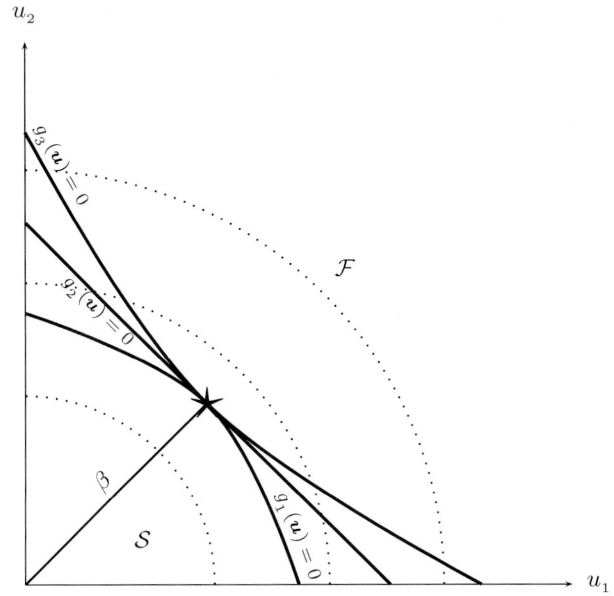

Fig. 7.2. Failure of the geometric reliability index to meet the ordering property requirement. The arcs indicate the joint density function. Different probability masses in the failure domain correspond to a single reliability index

it is desirable that the reliability indices would be $\beta_1 < \beta_2 < \beta_3 < \ldots$. The failure of the geometric index to meet this obvious requirement in the specific situation of mutually tangent limit state functions is illustrated in Fig. (7.2). This is due to the fact that the index is defined as a distance to the limit-state function and thus it is weakly related to it.

- *Positivity.* A property which has not been considered by researchers on this subject but which can be important for interpretation is that of positivity. This is a simple but essential condition because some degree of reliability should be expected from any safe domain however little it is. This condition is met by the widely used geometric index, as it is defined as the distance to the design point. Due to its wide application, it has generated a reasonable standard for interpretation of a reliability index, namely that it indicates a measure of the width of the safe domain in the standard space and, particularly, a distance measured from the origin to a critical point in general. However, the positiveness requirement is not always fulfilled by

the generalized index. This is an unavoidable consequence of its definition as a quantile of the standard Normal distribution, whose density symmetrically spreads over the negative and positive real axes. When the number of variables increases and the principal curvatures are negative the index becomes negative. A similar behavior is observed in the case of the empirical SORM index. This phenomenon could eventually create problems of interpretation in structural dynamics and stochastic finite element analyses, in which the dimensionality is very high. Such negativity is contrary to the standard interpretation promoted by the popular geometric index, which is evidently practical. For this reason, these indices are somewhat contradictory in that positive values could still be interpreted as distances but negative ones can only be interpreted in terms of a nominal probability of failure given by the standard Gaussian distribution. A unified framework for interpretation would be desirable, especially for code calibration and other applications.
As an illustration, Fig. 7.9 shows the generalized reliability index for the cylindrical problem

$$g(\boldsymbol{u}) = \eta^2 - \sum_{i=1}^{d} u_i^2 = 0 , \tag{7.10}$$

where u_i are standard Normal variables. The probability of failure is given by

$$P_{\mathrm{f}} = 1 - P[\chi_d^2 \leq \eta^2] , \tag{7.11}$$

where χ_d^2 denotes a Chi-square random variable with d degrees of freedom. This probability increases fast with d and, consequently, an index calculated as $\Phi^{-1}(P_{\mathrm{f}})$ becomes rapidly negative. This is just the expression of the generalized index for this case because $P[\boldsymbol{u} \in \mathcal{S}] = P[\chi_d^2 \leq \eta^2]$. Figure 7.9 was generated for $\eta = 3.5$ and it indicates that the generalized index becomes negative for $d > 11$. With respect to the empirical SORM index, however, it is not possible to determine its variation with the dimensionality in this case, because it is tied to the approximate SORM function (1.28) but, after the results shown in [188] it can be concluded that it becomes negative roughly as fast as the generalized index.

- *Dimension invariance.* By this expression it is meant the desirable sensitivity of the index to the dimensionality of the problem [40]. Using the same spherical example as above, it is evident that the geometric index is insensitive to the order of the sum, i.e. the number of dimensions, because $\beta = \eta$ for any d. The generalized and the empirical SORM indices correctly exhibit a sensitivity to d.
- *Easy computation.* The geometrical reliability index is obtained in a FORM analysis and hence it is difficult to compute in high dimensions with implicit limit state functions, as discussed in Chap. 1. On the other hand,

the generalized index substitutes the joint density function of the variables by the product of standard Gaussian marginals, so that its accurate computation requires a multi-dimensional integration, which for implicit limit-state functions implies a Monte Carlo simulation. For this reason, the structural reliability practice has favored the geometric index, despite the evident theoretical advantages of the generalized index. Finally, the use of the empirical SORM index requires the calculation of the principal curvatures, which for high dimensions is practically a very difficult task.

In the above discussion it has been shown that no index satisfies all the desirable properties.

Table 7.1. Comparison of hyperplane approximations $\langle \boldsymbol{w}, \boldsymbol{u} \rangle - b$ in structural reliability

Method	Equation	Characteristics
FORM	1.20	Gradient-based
		Minimization of distance to origin
		Very limited application to nonlinear functions
Fisher	4.13	Based on statistical moments of the population
		Maximization of class mean difference
		Applicable to nonlinear functions with large errors
Perceptron	4.16	Margin-based
		Minimization of misclassifications
		Applicable to nonlinear functions with large errors
SVM	5.10	Margin-based
		Maximization of margin
		Nonlinear cases treated as linear in a different space

7.3 A Comparison of Hyperplane Approximations

After the exposition of the various methodologies dealing with hyperplanes that have been derived in the fields of structural reliability and pattern recognition, we are in a position to compare several methods for hyperplane approximation of the limit-state function that are applicable to structural reliability. These are FORM, Fisher's criterion, Rosenblatt's Perceptron and Support Vector Machines (see Table 7.1).

FORM is based on Taylor expansion of the performance function and hence the hyperplane parameters are given in terms of the gradients of the performance function. Since normally these are not explicitly given they should be computed with finite-difference or perturbation techniques. Instead, the other three methods are based on samples of the performance function. Nevertheless, whereas in Rosenblatt's and Vapnik's methods the hyperplane parameters are directly determined by the samples and especially of those close to the boundary, in Fisher's criterion the samples are replaced by their global statistical measures, with the consequence that the information in the whereabouts of the boundary is lost. In this sense, the Perceptron and the SVM are closer to FORM than to Fisher's method, despite this later is also a pattern classification technique.

It is also interesting to note that all these methods involve an optimization solution. In FORM it is the constrained minimization of the distance to origin subject to the fact that the trial sample lies on the boundary between the classes. Fisher's method attempts to maximize the distance between the class-mean vectors, normalized by the pooled (or *within-class*) covariance matrix. In the Perceptron there is an unconstrained minimization of the classification errors formulated in terms of the functional margin, while in Support Vector Machines, the margin is maximized subject to correct classification.

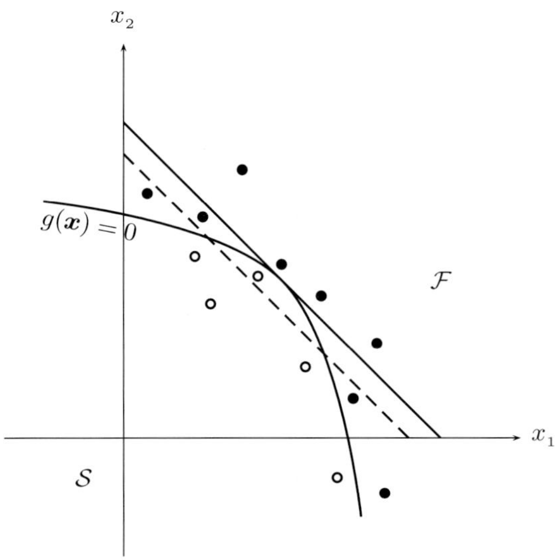

Fig. 7.3. Hyperplane methods applied to nonlinear limit states. While FORM draws a tangent hyperplane, as it is based on a Taylor expansion, Fisher, Rosenblatt and Vapnik (with a linear kernel) methods yield a secant hyperplane because they are driven by samples

The applicability of the first three hyperplane formulations to nonlinear cases is limited. In the FORM case there is the highest limitation, because the hyperplane is that of a tangent to the limit-state function, as it is based on the Taylor expansion of the performance function. On the contrary, Fisher's and Rosenblatt's methods (the latter when trained with the pocket instead of the original algorithm) can draw a secant hyperplane when applied to nonlinearly separated classes, as indicated by Fig. 7.3, because they are based on samples. By the same token, a Support Vector Machine does the same when trained with a linear kernel. Otherwise, it returs also a hyperplane but in a different space.

7.4 Secant Hyperplane Reliability Index

One of the shortcomings of the geometric reliability index is that it is based on a tangent hyperplane to the limit-state function. This is a consequence of its derivation using the Taylor expansion of the performance function, which evidently is a purely mathematical approach having no probabilistic considerations at all. Of course, the statistical measures of the basic variables are considered in the transformation from the input $\boldsymbol{x}-$space to the normalized $\boldsymbol{u}-$space. However, the probability content of the safe and failure domains are not explicitly incorporated into the formulation, as it is defined as a distance to a specific point.

It has been noted that this feature of the geometric index implies that the ordering property is not satisfied, and this was one of the main supports for the development of the generalized index [38]. However, the geometric index could be improved if it is no longer defined with respect to a tangent but to a secant hyperplane, which can be more meaningful from the engineering viewpoint, as has been remarked [40]. From the above discussion on reliability hyperplanes notice that in order to avoid the arbitrariness of such construction, explicit use should be made of the probabilistic information of the safe and failure domains, considered as statistical classes. This remedies the deficiency of the geometric index of not explicitly considering the probability masses lying on both sides of the limit-state function. In the absence of full probabilistic information of the basic variables, use can be made of the maximum entropy principle (see e.g. [91]) to invoke that the joint density function of the variables in the standard space is Gaussian, which is the correct decision from the point of view of Information Theory. Otherwise, if such information is available and it indicates that this is not the case, the transformation rules (Rosenblatt, Nataf or polynomial chaoses) can be used to produce such a model.

Accordingly, when only second-order statistical information is given, the multivariate Gaussian density can be used as a model for generating samples of the standardized random variables u_i that allow the calculation of a secant hyperplane. Hence the reliability index can be estimated as the distance to origin

$$\beta = \frac{b}{||\boldsymbol{w}||} \tag{7.12}$$

This is the basis of the first index proposed herein. The index could be construed with any of the methods illustrated in Fig. (7.3). However, Fisher's method leads to null indices. This is because the symmetry of the multivariate Gaussian density function which implies that the global mean and hence the offset b in (4.13) are null. If, on the other hand, use is made of a non centered population for enforcing a nonzero offset, then the maxentropic assumption would be violated. This leave us only with the Perceptron and the Support Vector methods are preferable in this respect.

Obviously, a hyperplane calculated by these two methods would not be useful for actual classification of samples in the generalization phase, because it is the result of a suboptimal model carrying out large training errors. However, these latter are but a manifestation of the curvatures of the limit state function, since the higher the errors, the lower the average curvature radius. In addition, the training errors are also a manifestation of the information lack linked to the second-order approximation to reliability, which could only be overcome if there is enough support for building a joint density function of the variables that allows leaving the maximum entropy supposition of a Gaussian density. In this sense, it can be said that the proposed secant hyperplane index is a trade-off solution to the reliability measurement problem that balances model complexity and available information. On the contrary, the geometric index can be regarded as a biased solution that is only admissible when the limit-state function is close to a plane.

7.4.1 Index Properties

Let us now show that an index calculated in this way satisfies the desirable properties:

- *Positivity.* Evidently, the index is positive as it is defined as a distance.
- *Globality and ordering property.* According to the discussion of the preceding section, Rosenblatt's Perceptron and Vapnik's Support Vector Machines with a linear kernel can be used for producing such a secant hyperplane. Since Rosenblatt's and Vapnik's methods are defined on the basis of training samples, they can easily produce a secant hyperplane that produces a satisfactory balance between the classification errors for both classes. This is due to the fact that the former responds to an unconstrained optimization in which the sum of misclassifications is minimized and the second applies the SRM inductive principle. These two approaches grant that the ordering property requirement is fulfilled. In the former case, seeking a minimal number of classification errors is equivalent to finding

a secant hyperplane that balances the safe and failure probability masses that are not correctly explained by the hyperplane, and this is exactly what is expected from a secant hyperplane from an engineering point of view. In the latter case, this same solution is achieved by a more sophisticated method which grants the best solution for a family of functions in terms of generalization abilities. Since both solutions are obtained from samples located in the entire input space, the solution is global in the sense defined above.

- *Formulation invariance.* Since the index is computed over classified samples, the formulation invariance requirement is satisfied, since the sign of e.g. $g(\boldsymbol{u})$ is the same of $g^3(\boldsymbol{u})$. In passing, this is a general characteristic of the classification view on the reliability problem.
- *Dimension invariance.* Let as consider again the problem (7.10). Since in this case the probability mass flows outwards as the dimensionality increases, the samples generated from it will flow accordingly and hence the secant hyperplane will move also outwards. Therefore, this important requirement is also fulfilled.
- *Easy computation.* The index can be easily calculated after rendering explicit the limit-state function using a flexible solver-surrogate technique, that is used to classify the samples.

In order to illustrate the actual fulfillment of the most relevant features, it suffices to contrive an example in which the dimension invariance is empirically demonstrated, because it subsumes the ordering property. This is because in problems showing an inward or outward probability flow parameterized with the dimensionality, a nested sequence of sets is in fact created by sequentially increasing the number of dimensions. Thus, the approximate SORM function (1.28) will be used [188]. It is repeated here for reference:

$$u_d = \beta + \frac{1}{2R} \sum_{i=1}^{d-1} u_i^2 \, . \tag{7.13}$$

Recall that β is the geometric (Hasofer-Lind's) reliability index, R is the average curvature radius and d the number of dimensions. The index was calculated for the values $R = 2, \beta = 2$ and for 2,3 and 4 dimensions. Notice that these sets are beyond the range of applicability of FORM and SORM, according to [187], due to the small radius. Therefore, the Empirical SORM index could not be calculated in this case. Contrary to the spherical problem (7.10), this problem is characterized by a decreasing probability mass in the failure domain as the dimensionality grows and, therefore, by an increasing reliability.

Figure 7.4 shows the training samples and the hyperplane obtained with Rosenblatt's Perceptron trained with the pocket algorithm [51] mentioned in Chap. 4. The departure from horizontality, determined by their calculation from samples instead of actual volumes, does not seriously affect its intercepts

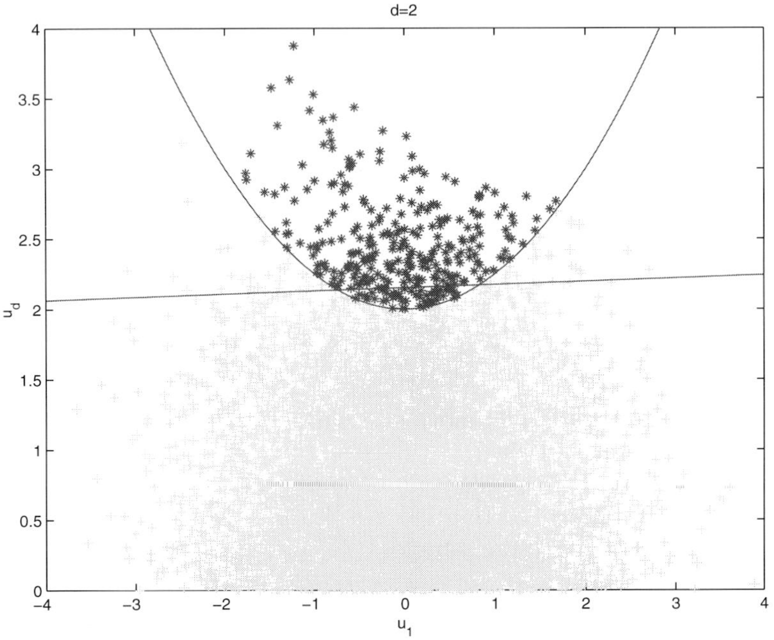

Fig. 7.4. Secant hyperplane classification index applied to a SORM function obtained with Rosenblatt's perceptron

with the axis u_d, which is the sought reliability index. This is shown in Table 7.2 for several dimensions. It can be observed that the secant hyperplanes move upwards with the dimensionality as expected. The fulfillment of the ordering and dimension invariance properties is evident.

The Table also shows the result obtained with Support Vector Machines with a linear kernel. It can be observed that the increase of the reliability indices with the dimensionality is more clearly marked in this case. Also it was observed that inclinations of the planes were much lower and the reliability indices among different runs were more stable than those given by the Perceptron. This latter fact is due to the optimality of the SVM hyperplane, which makes it more robust to sample perturbations than that obtained with the Perceptron algorithm [148]. Since the SVM technique also includes the minimization of the misclassifications in the non-separable case (see Chap. 6), it is the recommended method for deriving the secant hyperplane. In order to reflect the actual structure of the partition of the Gaussian hypersphere into the safe and unsafe domains, use should be made of a large number of

Gaussian samples. In practice, this implies that either the limit state function is explicit or it has been approximated by Neural Networks or nonlinear Support Vector Machines. The calculation of the Support Vector hyperplane over such a large number of samples is more practically carried out with efficient algorithms developed to this purpose [126, 93, 100].

Table 7.2. Hyperplane reliability indices of a SORM function

Dimensions	Geometric index	Secant index (Perceptron)	Secant index (SVM)
2	2	2.16	2.17
3	2	2.30	2.41
4	2	2.48	2.57

7.5 Volumetric Reliability Index

In the preceding section a secant reliability index that satisfies all the desirable properties was proposed on the basis of a classification regard to the structural reliability problem. In this section, it is shown that an alternative second-order reliability index that also fulfills the properties can be derived after a Support Vector approximation of the limit-state function.

Let us refer back to the discussion on the characteristics of indices proposed in the literature. First, note that the simple, geometric and generalized indices coincide in the specific case of a plane limit state. This coincidence indicates the special role played by hyperplane limit state functions. An alternative reliability index could hence be derived by converting any limit-state function, formulated in the standard space, into a hyperplane living in a different space using the Support Vector methodology. Hence, a geometrical measure of the hyperplane assuring all the desirable properties would give the reliability in such a transformed space. The nonlinear SV transformation should be such that a corresponding measure in the input space could be obtained. As shown in the construction of kernels in Chap. 3, this is assured in case of the polynomial kernel. Therefore, the index is defined by a restriction of the SV approximation of $g(\boldsymbol{u}) = 0$ with this kernel.

The advantages of a this proposed index are the following: First, it measures the reach of the limit-state function $g(\boldsymbol{u}) = 0$ within the entire domain of input variables, in a similar way to the generalized index, and not only of a point thereof, as the geometric index. Second, since the limit-state function is transformed into a hyperplane, the computation of geometrical quantities in the transformed space is very easy. Third, the only statistical information required is of second-order. Therefore, the index may be used as a measure of the reliability associated to a critical situation, because it is like a property of the limit-state function. More importantly, if sufficient information to characterize the density functions of the variables becomes available, the failure probability can be estimated on the very hyperplane on which the index is measured by Monte Carlo simulation. As a consequence the index can be regarded as a representative of the failure probability, which could eventually be assessed when there is sufficient information support. In this sense this volumetric index is stronger than the secant index introduced previously, which relies on the plausibility of the Gaussian assumption for generating the training samples on the basis of the maximum entropy principle. The volumetric index, on the contrary, does not rely any assumption of the kind.

Hence, it can be said that the SV methodology for classifier approximation allows dealing with information problem (which is a matter of concern in Structural Reliability, as discussed by [92, 115, 40, 43] and others) in a staged manner: (a) If only second-order information is available, a reliability index β can be derived by focusing on a geometrical description of the reach of the limit-state function in the safe domain by means of the proposed index; (b) If sufficient information becomes available that allows building an approximate joint density model of the basic variables, then a probability of failure P_{f} can be obtained in such a way that it uniquely corresponds to β because both have been obtained with the same SV hyperplane. This is not possible with the secant index proposed above, as it relies on the maximum entropy inference of the joint density function. On the other hand, under transformation of the basic variables to standard independent Normal variables, the generalized index is the only one among the classical indices that can be regarded as such a representative of the failure probability in the second-order information stage. However, the difficulties for its computation in the general case have hindered its application in practice.

Being a unique representative of the failure probability in the low-order information stage, the proposed index satisfies an intuitive hierarchy of the geometric and probabilistic approaches in structural reliability [43]: If the empirical information is scarce, the reliability should be assessed more in topological than in probabilistic terms; otherwise the purely probabilistic approach to reliability assessment is more justified (see also [40]).

7.5.1 Derivation of the Index

As was shown in Chap. 5, the support vector methodology is based on the high probability of correct hyperplane separation of two-class samples in a space of very high dimensionality. Assume that the input space is the standard space used on which the geometric index is defined. This is consistent with the assumption of scarce statistical information. Once the SVM hyperplane has been calculated, the new reliability index can be easily defined in the feature space and then transformed back to the input space.

The index is built by means of the SV approximation of the limit-state function strictly using the polynomial kernel for the reasons stated above. Since the hyperplane equation was normalized by the norm of the weight vector in deriving (5.5), the distance from the origin to the hyperplane in the feature space is actually given by

$$\bar{b} = \frac{b}{||\boldsymbol{w}||} . \tag{7.14}$$

Despite this value measures in some sense the reach of the safe domain, it is not a good basis for building a reliability index because it does not exhibit the ordering property. This is due to the fact that the positivity of the Lagrange multipliers α_i in (5.21) is altered by the values of the classes c_i, which are either $+1$ or -1 and hence there is no certainty that a bigger safe domain is characterized by a larger value of b. Therefore, it is necessary to devise a different index assuring the ordering property. This is simply the volume determined by the hyperplane because the volume of the safe domain is a natural measure of the reliability satisfying the ordering property [40]. Although, since the safe domain is normally unbound (a classical exemption is the problem of a column subject to axial compression and flexural moments), it is necessary to extract a standard measure from this infinite volume. This is not easy for general nonlinear limit-states but in this case it is facilitated by the very geometric nature of the hyperplane, as it happens to cross every axis in a single point. The intercepts occur in a sub-domain that will be called the fundamental orthant.

With reference to (5.14) observe that $\zeta_i = u_i^2 > 0, \ i = 1, 2, \ldots, d$ so that the useful domain is restricted to the orthants with $\zeta_i > 0$ since otherwise the pre-images in the input standard space would be imaginary. Now, the rest $\zeta_k = \sqrt{2} u_i u_j, \ k > d, \forall i, j$ can be either negative or positive. Hence, the safe domain is unbound only along these latter coordinates. The fundamental volume and its extension along a coordinate of the crossed type is shown in Fig. 7.5. The intercepts of the hyperplane in the fundamental orthant are given by

$$\bar{\zeta_i} = \frac{\bar{b}}{w_i} . \tag{7.15}$$

They define a $D-$dimensional hyper-tetrahedron having a vertex on the origin, where D is the dimensionality of the feature space, according to the nomenclature adopted in Chaps. 3 and 5. This means that the fundamental volume is

$$V_{\scriptscriptstyle 0} = \frac{1}{D!} \prod_{i=1}^{D} |\bar{\zeta_i}| , \tag{7.16}$$

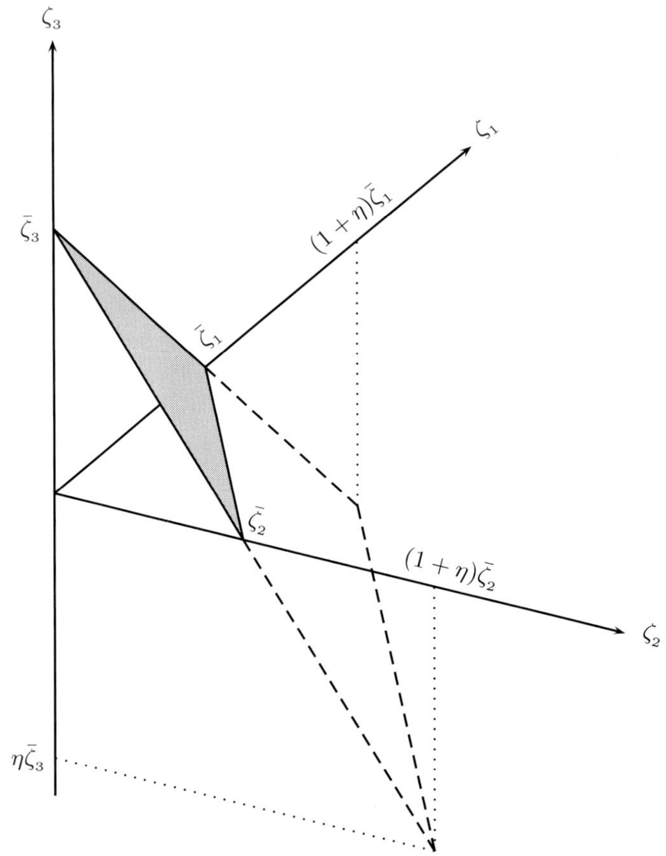

Fig. 7.5. Extension of the fundamental volume for the definition of the reliability index

where the absolute value has been introduced for assuring the positivity of the volume, because some intercepts on axes of the cross type can be negative as said above. Since any portion of the infinite volume can be expressed in terms of it in a simple linear form, the volume V_0 is a standard measure of the reach of the limit state function in the safe domain. A more meaningful standard, however, is given by the equivalent hypercube size

$$B = V_0^{1/D} , \tag{7.17}$$

which is back mapped from the feature space onto the input space, where it has an engineering meaning. This explains why the polynomial kernel was selected to this derivation, as it allows the transformation from any distance B in the feature space to the corresponding value in the input space. It can be easily shown that this mapping is given by

$$\beta = \sqrt{B^{2/p} - \theta}\ , \tag{7.18}$$

where θ is the bias in the polynomial kernel (see Table 3.1). Hence, for $\theta = 0$ the reliability index is

$$\beta = \sqrt{V_0^{2/pD}}\ . \tag{7.19}$$

7.5.2 Index Properties

Let us now show that the index fulfills all the desirable properties:

1. *Formulation invariance.* Similarly to the secant index, this property is also fulfilled in this case because it is based on a classification approach.
2. *Global measure.* The proposed index is not tied to one or more specific points of the limit state function and hence it can be interpreted as a global measure of the reliability. This property is also exhibited by the generalized index, but not by the geometric index. Notice that this new reliability index is uniquely defined for the hyperplane in the feature space, up to the proximity of the support vectors to the limit state function. Accordingly, the problem of multiple candidates to design points for a single limit state function in the input space, which occurs frequently in practice when using the geometric reliability index, is avoided.
3. *Ordering property.* The index also exhibits this important property. This is due to the fact that it is defined in terms of the volume of the hypertetrahedron, which grows geometrically as the position of the samples defining the limit state are the farther from the origin in the input space, as illustrated by Fig. 7.6. The index, hence, measures the amount of uncertainty on the basic variables, as measured by second order moments, that the limit state function can accommodate without failure. Notice that the generalized index (7.7) is equivalent to compute the weighted volume of the same domain using the multivariate independent Gaussian density as weighting function, whose effect is removed by applying the inverse distribution function. The weight was introduced in such proposal in order to avoid dealing with improper integrals. In the present proposal, the volume of the hyper-tetrahedron in the orthant where all the intercepts occur is taken as the reference volume defining the reliability measure on the basis of the linearity of the hyperplane. As a consequence, no weighting function is needed to compute the reference volume and hence the inverse operation is not required.

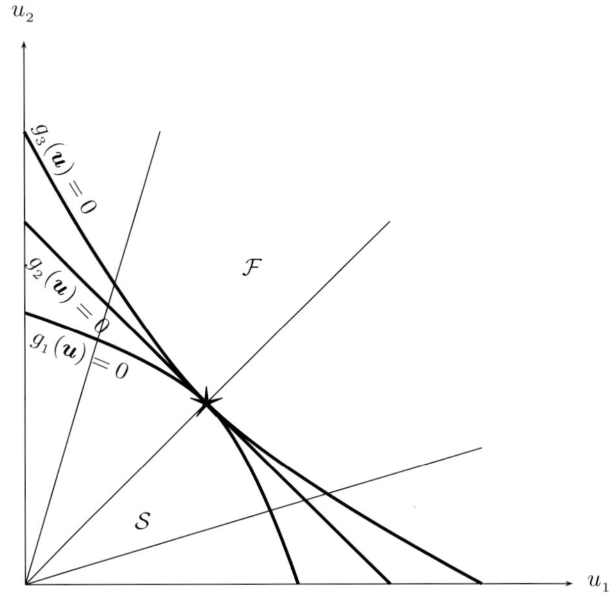

Fig. 7.6. On the ordering property of the volumetric index

In order to demonstrate the fulfillment of this essential property, a two-dimensional example is useful as it facilitates the visualization of the feature space. The limit-state functions are three parabolas defined as

$$
\begin{aligned}
g_1(u_1, u_2) &= 3 - \frac{u_1^2}{3} - u_2 \,, \\
g_3(u_1, u_2) &= 4 - \frac{u_1^2}{4} - u_2 \,, \\
g_4(u_1, u_2) &= 5 - \frac{u_1^2}{5} - u_2 \,,
\end{aligned}
\tag{7.20}
$$

and a plane

$$
g_2(u_1, u_2) = -1.464u_1 - u_2 + 4.605 \tag{7.21}
$$

which is tangent to the first parabola at the design point (see Fig. 7.7). To these functions there correspond safe sets forming a nested structure

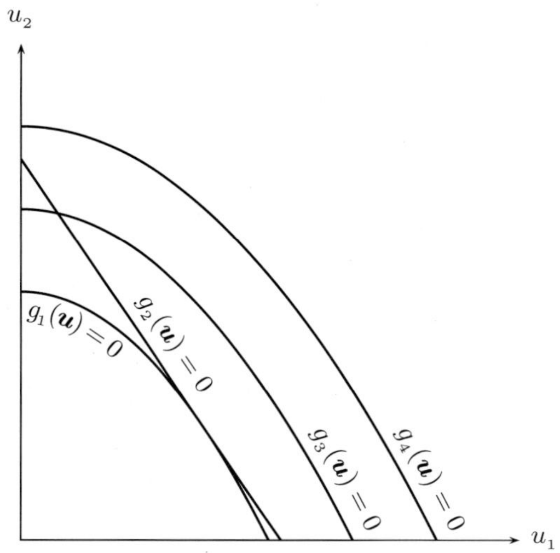

Fig. 7.7. Nested set of safe domains used for showing the fulfillment of the ordering property by the SVM-based reliability index. Note that function $g_2(\boldsymbol{u}) = 0$ is tangent to $g_1(\boldsymbol{u}) = 0$ and hence they have the same geometric index

$\mathcal{S}_1 \subset \mathcal{S}_2 \subset \mathcal{S}_3 \subset \mathcal{S}_4$. Therefore, it is desirable that the reliability indices satisfy the inequality chain $\beta_1 < \beta_2 < \beta_3 < \beta_4$.

Table 7.3. Reliability indices

Function	Geometric index	Volumetric index
$g_1(\boldsymbol{u})$	2.59	2.55
$g_2(\boldsymbol{u})$	2.59	2.66
$g_3(\boldsymbol{u})$	3.46	3.41
$g_4(\boldsymbol{u})$	4.33	4.25

Figure 7.8 is a scaled representation of the hyperplanes in feature space corresponding to g_1 and g_2 obtained with $\theta = 0$ and $p = 2$. The crossing

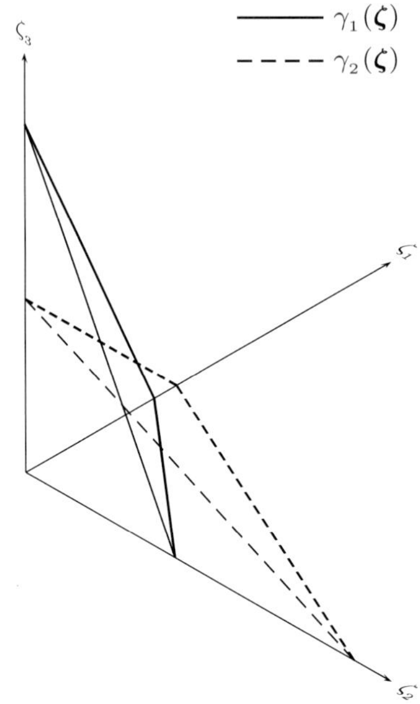

Fig. 7.8. Feature-space hyperplanes equivalent to the nested limits-state functions. The solid line corresponds to $g_1(\boldsymbol{u})$ and the dashed line to $g_2(\boldsymbol{u})$. The crossing of the hyperplanes is a reflect in the feature space of their contact in the input space

of the two planes is a reflect of the tangency of the latter at the design point. Table 7.3 shows the β values obtained with (7.19). The Table also displays the values of the geometric index for comparison. Notice that at a difference to this latter, which yields $\beta_1 = \beta_2$, the proposed index satisfies the desired ordering property. In addition, note that the values for g_1 and g_2 bracket the single result of the geometric index and that there is a remarkable similarity of the two indices for all the functions, so that the proposed index is not at odds with the customary values of the structural reliability practice.

4. *Positivity.* This condition is obviously met by the geometric index, as it is defined as a distance. However, the generalized index fails to meet this requirement for negative curvatures and a moderate number of dimensions. On the other hand, the empirical SORM index has a similar behavior, as illustrated by several examples in [188]. The negativity of the generalized and empirical SORM indices creates problems of interpretation as already

said. In particular, the generalized index is intended to measure the reach of the limit-state function in the safe domain and this is in conflict with a negative value.

This phenomenon has a link to the curse of dimensionality, whose main features were summarized in Chap. 2. In fact, as the dimensionality increases, the mean Euclidean distance to the origin also increases [9]. This increase is, of course, modeled by the multidimensional probability density of the samples. Therefore, the negativity of the mentioned indices for high dimensions is a consequence of the involvement of the very mathematical expression of the curse of dimensionality in their definition. In contrast, since the proposed index is defined only in terms of a plain volume without resort to probability functions, it is always positive.

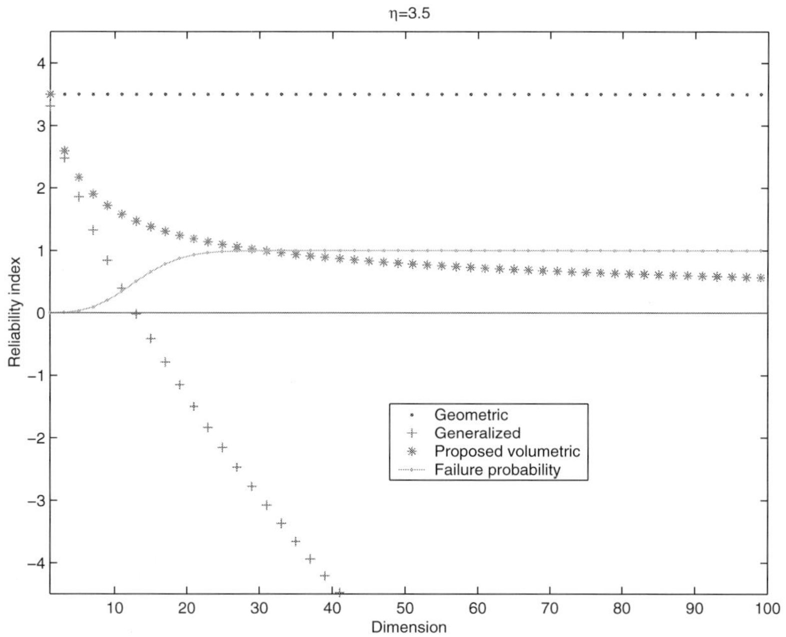

Fig. 7.9. Reliability indices and the dimension invariance property

5. *Dimension invariance.*
   In order to illustrate that the proposed volumetric index fulfills this property, Fig. 7.9 shows the dimensionality dependence of the geometric, generalized and proposed indices for the spherical problem (7.10) for $\eta = 3.5$. The failure probability is also depicted in the figure. The failure of the geometric index to fulfill the dimension invariance requirement is obvious. More interesting is to note that the proposed index follows a slowly decaying trend that agrees with the also slow increase of the failure probability after the point of crossing of the two graphs, which happens approximately at the point where the probability is close to its maximum. This is contrast to the generalized index, whose continuously decreasing trend implies that its negative values correlate better with the logarithm of the complement of the said probability, as it progressively loses the distance meaning that could be attributed to positive ones. Notice also that the proposed index crosses the curve of the failure probability.
6. *Easy computation.* Obviously, a hypercube equivalent to the safe domain could be calculated without the nonlinear transformation to the feature space performed by the support vector methodology. However, this would imply the calculation of the actual volume of the domain, which is a task that for several dimensions and complex hypersurfaces requires a Monte Carlo simulation, as is well known. If, in addition, the limit state function is given implicitly, as is commonly the case in the reliability analysis of structural systems, then each simulation must be performed with the aid of the numerical solver of the structure, thereby increasing the computational labor. These difficulties are shared by the generalized index, as it substitutes the (possibly unknown) joint density function of the variables by the product of standard Gaussian marginals, so that its accurate computation requires a Monte Carlo simulation in the general case. On the contrary, in the proposed approach the volume is calculated with a simple formula in the feature space. Besides, the problem of implicit limit state functions is solved with parsimonious algorithms exploiting the key instrument of margins, such as that exposed in Chap. 5.

References

1. M. A. Aizerman, E. M. Braverman, and L. I. Rozonoer. Theoretical foundations of the potential function method in pattern recognition. *Automation and Remote Control*, 25:821 – 837, 1964.
2. M. A. Aizerman, E. M. Braverman, and L. I. Rozonoer. The robbins-monroe process and the method of potential functions. *Automation and Remote Control*, 28:1992 – 1885, 1965.
3. G. L. Ang, A. H-S. Ang, and W. H. Tang. Optimal importance sampling density estimator. *Journal of Engineering Mechanics*, 118:1146 – 1163, 1991.
4. M. Anthony and P. L. Bartlett. *Neural Network Learning: Theoretical Foundations*. Cambridge University Press, Cambridge, 1999.
5. G. A. Athanassoulis and P. N. Graviladis. The truncated hausdorff moment problem solved with kernel density functions. *Probabilistic Engineering Mechanics*, 17:273 – 291, 2002.
6. K. E. Atkinson. *The Numerical Solution of Integral Equations of the Second Kind*. Cambridge University Press, Cambridge, 1997.
7. S. K. Au and J. L. Beck. A new adaptive importance sampling scheme for reliability calculations. *Structural Safety*, 21:135 – 158, 1999.
8. S. K. Au and J. L. Beck. Estimation of small failure probabilites in high dimensions by subset simulation. *Probabilistic Engineering Mechanics*, 16:263 – 277, 2001.
9. S. K. Au and J. L. Beck. Importance sampling in high dimensions. *Structural Safety*, 25:139 – 163, 2003.
10. T. Bäck. Evolution strategies: An alternative evolutionary algorithm. In J.M. Alliot, E. Lutton, E. Ronald, M. Shoenhauer, and D. Snyers, editors, *Artificial Evolution*, pages 3–20. Springer-Verlag, Berlin, 1996.
11. T. Bäck, F. Hoffmeister, and H. P. Schwefel. A survey on evolution strategies. In R.K. Belew and L.B. Booker, editors, *Proceedings of the Fourth International Conference on Genetic Algorithms*, pages 2–9, San Mateo, California, 1991. Morgan Kauffman Publishers.
12. A.R. Barron. Universal approximation bounds for superpositions of a sigmoidal function. *IEEE Trans. Inform. Theory*, IT-39:930, 1993.
13. R. Bellman. *Adaptive Control Processes: A Guided Tour*. Princenton University Press, New Jersey, 1961.

14. Y. Ben-Haim. *Robust Reliability in the Mechanical Sciences.* Springer Verlag, Berlin, 1996.
15. J. R. Benjamin and C. A. Cornell. *Probability, Statistics and Decision for Civil Engineers.* McGraw-Hill, New York, 1970.
16. C. M. Bishop. *Neural Networks for Pattern Recognition.* Oxford University Press, Oxford, 1995.
17. P. Bjerager. Methods for structural reliability computation. In F. Casciati and J. B. Roberts, editors, *Reliability Problems: General principles and Applications in Mechanics of Solids and Structures*, pages 89–136. Springer-Verlag, Wien, 1991.
18. G. E. P. Box and N. R. Draper. *Empirical Model Building and Response Surfaces.* John Wiley and Sons, New York, 1987.
19. L. Breiman. Bagging predictors. *Machine Learning*, 26:126 – 140, 1996.
20. L. Breiman, J. H. Friedman, R. A. Olshen, and C. J. Stone. *Classification and Regression Trees.* Wadsworth Inc., New York, 1984.
21. C. Bucher. A fast and efficient response surface approach for structural reliability problems. *Structural Safety*, 7:57–66, 1990.
22. C. G. Bucher. Adaptive Sampling: An Iterative Fast Monte-Carlo Procedure. *Structural Safety*, 5:119 – 126., 1988.
23. C. G. Bucher and Y. K. Lin. Stochastic stability of bridges considering coupled modes. *Journal of Engineering Mechanics*, 114:2055 – 2071, 1988.
24. C. G. Bucher, H. J. Pradlwarter, and G. I. Schuëller. Computational stochastic structural analysis. In G. I. Schuëller, editor, *Structural Dynamics: Recent Advances*, pages 285 – 300. Springer-Verlag, Berlin, 1991.
25. S. V. S. Cabral and L. S. Katafygiotis. Neural network based response surface method and adaptive importance sampling for reliability analysis of large structural systems. In R. B. Corotis, G. I. Schuëller, and M. Shinozuka, editors, *Structural Safety and Reliability - Proceedings of the International Conference on Structural Safety and Reliability, ICOSSAR 01, Newport Beach, California, 17-22 June 2001*, page 46, Lisse, The Netherlands, 2001. A.A.Balkema, Publishers.
26. O. J. Chapman and A. D. Crossland. Neural networks in probabilistic structural mechanics. In C. (Raj) Sundararajan, editor, *Probabilistic Structural Mechanics Handbook*, pages 317–330. Chapman & Hall, New York, 1995.
27. V. Cherkassky and F. Mulier. *Learning from Data.* John Wiley and Sons, New York, 1998.
28. A. Cichocki and R. Unbehauen. *Neural Networks for Optimization and Signal Processing.* John Wiley and Sons, Chichester, 1993.
29. C. A. Cornell. A probability-based structural code. *Journal of the American Concrete Institute*, 66:974 – 985., 1969.
30. C. Cortes and V. N. Vapnik. Support vector networks. *Machine Learning*, 20:273 – 297, 1995.
31. T. M. Cover. Geometrical and statistical properties of systems of linear inequalities with applications in pattern recognition. *IEEE Transactions on Electronic Computers*, 14:326 – 334, 1965.
32. N. A. C. Cressie. *Statistics for Spatial Data.* John Wiley and Sons, New York, 1993.
33. N. Cristianini and J. Shawe-Taylor. *An Introduction to Support Vector Machines and other Kernel-based Methods.* Cambridge University Press, Cambridge, 2000.

34. A. P. Dempster, N. M. Laird, and D. B. Rubin. Maximum likelihood from incomplete data via the em algorithm. *Journal of the Royal Statistical Society, B*, 39:1 – 38, 1977.
35. A. DerKiureghian, H. Z. Lin, and S. J. Hwang. Second-order reliability approximations. *Journal of Engineering Mechanics*, 113:1208–1225, 1987.
36. A. DerKiureghian and P. L. Liu. Structural reliability under incomplete probability information. *Journal of Engineering Mechanics*, 112:85–104, 1986.
37. K. I. Diamantaras. Neural networks and principal component analysis. In Y. H. Hu and J. N. Hwang, editors, *Handbook of Neural Network Signal Processing*. C.R.C. Press, Boca Ratón, 2001.
38. O. Ditlevsen. Generalized second moment reliability index. *Journal of Structural Mechanics*, 7:435–451, 1979.
39. O. Ditlevsen. Principle of Normal tail approximation. *Journal of the Engineering Mechanics Division*, 107:1191 – 1207, 1981.
40. O. Ditlevsen and H. O. Madsen. *Structural Reliability Methods*. John Wiley and Sons, Chichester, 1996.
41. I. Doltsinis and F. Rau. On the ordinary distance between multivariate random systems in engineering. *Computer Methods in Applied Mechanics and Engineering*, 191:133 – 156, 2001.
42. R. O. Duda, P. E. Hart, and D. G. Stork. *Pattern Classification*. John Wiley and Sons, New York, 2001.
43. I. Elishakoff. Essay on reliability index, probabilistic interpetation of safety factor and convex models of uncertainty. In F. Casciati and J. B. Roberts, editors, *Reliability Problems: General principles and Applications in Mechanics of Solids and Structures*, pages 237–271. Springer-Verlag, Wien, 1991.
44. G. K. Er. A method for multi-parameter PDF estimation of random variables. *Structural Safety*, 20:25 – 36, 1998.
45. L. Faravelli. Response-surface approach for reliability analysis. *Journal of the Engineering Mechanics*, 115:2763 – 2781, 1989.
46. R. V. Field and M. Grigoriu. On the accuracy of the polynomial chaos approximation. *Probabilistic Engineering Mechanics*, 19:65 – 80, 2004.
47. T. Fine. *Feedforward Neural Network Methodology*. Springer Verlag, New York, 1999.
48. R. A. Fisher. *Contributions to Mathematical Statistics*. John Wiley and Sons, New York, 1952.
49. A. Florian. An efficient sampling scheme: Updated Latin Hypercube Sampling. *Probabilistic Engineering Mechanics*, 7:123 – 130, 1992.
50. Y. Freund and R. Schapire. A decision-theoretic generalization of on-line learning and an application to boosting. *Journal of Computer and System Sciences*, 55:119 – 139, 1997.
51. S. I. Gallant. Perceptron based learning algorithms. *IEEE Transactions on Neural Networks*, 1:179 – 191, 1990.
52. R. Ghanem. Stochastic finite elements with multiple random non-Gaussian properties. *Journal of Engineering Mechanics*, 125:26–40, 1999.
53. R. G. Ghanem. Hybrid stochastic finite elements and generalized monte carlo simulation. *ASME Journal of Applied Mechanics*, 65:1004 – 1009, 1998.
54. R. G. Ghanem and R. M. Kruger. Numerical solution of spectral stochastic finite element systems. *Computer Methods in Applied Mechanics and Engineering*, 129:289 – 303, 1996.

55. R. G. Ghanem and P. D. Spanos. *Stochastic Finite Elements: A Spectral Approach.* Springer Verlag, New York, 1991.
56. R. G. Ghanem and P. D. Spanos. Spectral techniques for stochastic finite elements. *Archives of Computational Methods in Engineering*, 4:63 – 100, 1997.
57. F. Girosi and T. Poggio. Networks and the best approximation property. *Biological Cybernetics*, 63:169 – 176, 1990.
58. D.E. Goldberg. *Genetic Algorithms in search, optimization and machine learning.* Addison-Wesley, New York, 1989.
59. H. M. Gomes and A. M. Awruch. Comparison of response surface and neural network with other methods for structural reliability analysis. *Structural Safety*, 26:49–67, 2004.
60. M. Grigoriu. *Stochastic Calculus. Applications in Science and Engineering.* Birkhäuser, Boston, 2002.
61. X. L. Guan and R.E. Melchers. Effect of response surface parameter variation on structural reliability estimates. *Structural Safety*, 23:429 – 444, 2001.
62. J. Hadamard. *Lectures on the Cauchy problem in linear partial differential equations.* Yale University Press, New Haven, 1923.
63. A. Haldar and S. Mahadevan. *Reliability Assessment using Stochastic Finite Element Analysis.* John Wiley and Sons, New York, 2000.
64. A. Harbitz. An efficient sampling method for probability of failure calculation. *Structural Safety*, 3:109 – 115., 1986.
65. N. Harnpornchai, H. J. Pradlwarter, and G. I. Schuëller. Stochastic analysis of dynamical systems by phase-space controlled Monte Carlo simulation. *Computer Methods in Applied Mechanics and Engineering*, 168:273–283, 1999.
66. E. J. Hartman, J. D. Keeler, and J. M. Kowalski. Layered neural networks with gaussian hidden units as universal approximations. *Neural Computation*, 2:210 – 215, 1990.
67. A. M. Hasofer. Non-parametric estimation of failure probabilities. In F. Casciati and B. Roberts, editors, *Mathematical Models for Structural Reliability Analysis*, pages 195–226. CRC Press, Boca Ratón, 1996.
68. A. M. Hasofer and N. C. Lind. Exact and invariant second moment code format. *Journal of the Engineering Mechanics Division*, 100:111 – 121, 1974.
69. T. Hastie, R. Tibshirani, and J. Friedman. *The Elements of Statistical Learning.* Springer Verlag, New York, 2001.
70. R. Herbrich. *Learning Kernel Classifiers.* The M. I. T. Press, Cambridge, 2002.
71. R. Herbrich, T. Graepel, and C. Campbell. Bayes point machines. *Journal of Machine Learning Research*, 1:245 – 279, 2001.
72. D. B. Hernández. *Lectures on Probability and Second Order Random Fields.* World Scientific, Singapore, 1995.
73. M. Hohenblicher and R. Rackwitz. Non-normal dependent vectors in structural reliability. *Journal of the Engineering Mechanics Division*, 107:1127 – 1138, 1981.
74. H. P. Hong and N. C. Lind. Approximate reliability analysis using normal polynomial and simulation results. *Structural Safety*, 18:329 – 339., 1996.
75. K. M. Hornik, M. Stinchcommbe, and H. White. Multilayer feedforward networks are universal approximators. *Neural Networks*, 2:359 – 366, 1989.
76. D.E. Huntington and C.S. Lyrintzis. Improvements to and limitations of Latin Hypercube Sampling. *Probabilistic Engineering Mechanics*, 13:245 – 253, 1998.
77. J. E. Hurtado. Neural networks in stochastic mechanics. *Archives of Computational Methods in Engineering*, 8:303 – 342, 2001.

78. J. E. Hurtado. Analysis of one-dimensional stochastic finite elements using neural networks. *Probabilistic Engineering Mechanics*, 17:35 – 44, 2002.
79. J. E. Hurtado and D. A. Alvarez. Neural network-based reliability analysis: A comparative study. *Computer Methods in Applied Mechanics and Engineering*, 191:113 – 132, 2001.
80. J. E. Hurtado and D. A. Alvarez. A classification approach for reliability analysis with stochastic finite element modeling. *Journal of Structural Engineering*, 129:1141 – 1149, 2003.
81. J. E. Hurtado, D. A. Alvarez, and A. H. Barbat. Monte Carlo analysis of structural systems using neural networks. In G.I. Schuëller and P.D. Spanos, editors, *Monte Carlo Simulation - Proceedings of the International Conference on Monte Carlo simulation, MCS-2000, Monte Carlo, Monaco, 18-21 June 2000*, pages 265–271, Lisse, The Netherlands, 2000. A.A.Balkema, Publishers.
82. J. E. Hurtado and A. H. Barbat. Analysis of stochastic finite elements using neural classifiers. In R. B. Corotis, G. I. Schuëller, and M. Shinozuka, editors, *Structural Safety and Reliability - Proceedings of the International Conference on Structural Safety and Reliability, ICOSSAR 01, Newport Beach, California, 17-22 June 2001*, page 132, Lisse, The Netherlands, 2001. A.A.Balkema, Publishers.
83. J. E. Hurtado and A.H. Barbat. Fourier-based maximum entropy method in stochastic dynamics. *Structural Safety*, 20:221 – 235, 1998.
84. J. E. Hurtado, J. M. Londono, and M. A. Meza. On the applicability of neural networks for soil dynamics amplification analysis. *Soil Dynamics and Earthquake Engineering*, 21:619 – 631, 2001.
85. R. A. Ibrahim. *Parametric Random Vibration.* Research Studies Press Ltd., Taunton, 1985.
86. R. L. Iman and W. J. Canover. Small sample sensitivity analysis techniques for computer models with an application to risk assessment. *Communications in Statistics, Theory and Methods*, A9:1749 – 1842., 1980.
87. H. Ishibuchi and H. Tanaka. Fuzzy regression analysis using neural networks. *Fuzzy Sets and Systems*, 50:257 – 265, 1987.
88. J. S. R. Jang, C. T. Sun, and E. Mizutani. *Neuro Fuzzy and Soft Computing.* Prentice Hall, Upper Saddle River, USA, 1997.
89. P. Kall and S. W. Wallace. *Stochastic Programming.* John Wiley and Sons, Chichester, 1995.
90. G. Kallianpur. *Stochastic Filtering Theory.* Springer Verlag, New York, 1980.
91. J. N. Kapur. *Maximum Entropy Models in Science and Engineering.* John Wiley and Sons, New York, 1989.
92. D. Kececioglu and G. Lamarre. Designing mechanical components to a specified reliability and confidence level. *Nuclear Engineering and Design*, 50:149 – 162, 1978.
93. S. S. Keerthi, S. K. Shevade, C. Bhattacharyya, and K. R. K. Murthy. A fast iterative nearest point algorithm for support vector machine classifier design. Technical report, Intelligent Systems Lab, Department of Computer Science and Automation, Indian Institute of Science, 1999.
94. C. A. Kennedy and W. C. Lennox. Solution to the practical problem of moments using non-classical orthogonal polynomials with applications for probabilistic analysis. *Probabilistic Engineering Mechanics*, 15:371 – 379, 2000.

95. C. A. Kennedy and W. C. Lennox. Moment operations on random variables, with applications for probabilistic analysis. *Probabilistic Engineering Mechanics*, 16:253 – 259, 2001.
96. S. H. Kim and S. W. Na. Response surface method using vector projected sampling points. *Structural Safety*, 19:3–19, 1997.
97. M. Kleiber and T. D. Hien. *The Stochastic Finite Element Method.* John Wiley and Sons, Chichester, 1992.
98. T. Kohonen. *Self-Organizing Maps.* Springer, Berlin, 1997.
99. S. Kotz, N. Balakrishnan, and N. L. Johnson. *Continuous Multivariate Distributions, Vol. 1.* John Wiley and Sons, New York, 2000.
100. A. Kowalczyk. Maximal margin perceptron. In A. Smola, P. Bartlett, B. Schölkopf, and D. Schuurmans, editors, *Advances in Large Margin Classifiers*, pages 75 – 113. The M. I. T. Press, Cambridge, 2000.
101. Y. LeCun. Learning processes in an asymmetric threshold network. In *Disordered Systems and Biological Organizations*, pages 233–240, Les Houches, 1986. Springer.
102. Y. K. Lin and C. Q. Cai. *Probabilistic Structural Dynamics. Advanced Theory and Applications.* McGraw-Hill, New York, 1995.
103. J. S. Liu. *Monte Carlo Strategies in Scientific Computing.* Springer, New York, 2001.
104. P. L. Liu and A. DerKiureghian. Multivariate distribution models with prescribed marginals and covariances. *Probabilistic Engineering Mechanics*, 1:105 – 112, 1986.
105. W. K. Liu, T. Belytschko, and Y. J. Lua. Probabilistic finite element method. In C. (Raj) Sundararajan, editor, *Probabilistic Structural Mechanics Handbook*, pages 70–105. Chapman & Hall, New York, 1995.
106. I. D. Lonstaff and J. F. Cross. A pattern recognition approach to understanding the multi-layer perceptron. *Pattern Recognition Letters*, 5:315 – 319, 1987.
107. F. L. Luo and R. Unbehauen. *Applied Neural Networks for Signal Processing.* Cambridge University Press, Cambridge, 1998.
108. S. Mallat. *A Wavelet Tour of Signal Processing.* Academic Press, San Diego, 1999.
109. W. L. Martinez and A. R. Martinez. *Computational Statistics Handbook with MATLAB.* Chapman & Hall / CRC, Boca Ratón, 2002.
110. L. Mason, J. Baxter, P. L. Bartlett, and M. Frean. Functional gradient techniques for combining classifiers. In A. Smola, P. Bartlett, B. Schölkopf, and D. Schuurmans, editors, *Advances in Large Margin Classifiers*, pages 221 – 246. The M. I. T. Press, Cambridge, 2000.
111. G. McLachlan and D. Peel. *Finite Mixture Models.* John Wiley and Sons, New York, 2000.
112. R. E. Melchers. Importance sampling in structural systems. *Structural Safety*, 9:117 – 128, 1990.
113. R. E. Melchers. *Structural Reliability: Analysis and Prediction.* John Wiley and Sons, Chichester, 1999.
114. P. G. Melnik-Melnikov and E. S. Dekhtyaruk. Rare events probabilities estimation by "Russian Roulette and Splitting" simulation technique. *Probabilistic Engineering Mechanics*, 15:125 – 129, 2000.
115. J. Menčik. Reliability assessment with small amounts of experimental data. In R. B. Corotis, G. I. Schuëller, and M. Shinozuka, editors, *Structural Safety and*

Reliability - Proceedings of the International Conference on Structural Safety and Reliability, ICOSSAR 01, Newport Beach, California, 17-22 June 2001, page 187, Lisse, The Netherlands, 2001. A.A.Balkema, Publishers.
116. N. Metropolis, A. W. Rosenbluth, M. N. Rosenbluth, A. H. Teller, and E. Teller. A new algorithm for adaptive multidimensional integration. *Journal of Chemical Physics*, 21:1087 – 1092, 1953.
117. H. Mhaskar. Aproximation properties of a multilayered feedforward artificial neural network. *Advances in Computational Mathematics*, 1:61 – 80, 1993.
118. J. C. Mitteau. Error evaluations for the computation of failure probability in static structural reliability problems. *Probabilistic Engineering Mechanics*, 14:119 – 135, 1999.
119. D. C. Montgomery. *Design and Analysis of Experiments*. John Wiley and Sons, New York, 1991.
120. A. Nataf. Determination des distributions dont les marges sont donées. *Comptes rendues de l'Academie des Sciences*, 225:42–43, 1962.
121. A. B. J. Novikoff. On convergence proofs of perceptrons. In *Proceedings of the Symposium on the Mathematical Theory of Automata*, volume XII, pages 615–622, New York, 1986. Polytechnic Institute of Brooklyn.
122. A. Olsson, G. Sandberg, and O. Dahlblom. On Latin Hypercube sampling for structural reliability analysis. *Structural Safety*, 25:47–68, 2003.
123. M. Papadrakakis, V. Papadopoulos, and N.D. Lagaros. Structural reliability analysis of elastic-plastic structures using neural networks and Monte Carlo simulation. *Computer Methods in Applied Mechanics and Engineering*, 136:145 – 163, 1996.
124. E. Parzen. On estimation of probability density function and mode. *Annals of Mathematical Statistics*, 33:1065 – 1076, 1962.
125. K. K. Phoon, S. P. Huang, and S. T. Quek. Implementation of Karhunen-Loeve expansion for simulation using a wavelet-Galerkin scheme. *Probabilistic Engineering Mechanics*, 17:293 – 303, 2002.
126. J. C. Platt. Fast training of support vector machines using sequential minimal optimization. In B. Schölkopf, C. J. C. Burges, and A. Smola, editors, *Advances in Kernel Methods*, pages 185 – 208. The M. I. T. Press, Cambridge, 1999.
127. D. S. G. Pollock. *A Handbook of Time-Series Analysis, Signal Processing and Dynamics*. Academic Press, San Diego, 1999.
128. K. Popper. *The Logic of Scientific Discovery*. Harper Torch Book, New York, 1968.
129. H. J. Pradlwarter. Deterministic integration algorithms for stochastic response computations of FE-systems. *Computers and Structures*, 80:1489–1502, 2002.
130. H. J. Pradlwarter, C. G. Bucher, and G. I. Schuëller. Nonlinear systems. In G. I. Schuëller, editor, *Structural Dynamics: Recent Advances*, pages 128 – 213. Springer-Verlag, Berlin, 1991.
131. H. J. Pradlwarter and G. I. Schuëller. Assessment of low probability events of dynamical systems by controlled Monte Carlo simulation. *Probabilistic Engineering Mechanics*, 14:213 – 227, 1999.
132. H. J. Pradlwarter, G. I. Schuëller, and P. G. Melnik-Melnikov. Reliability of MDOF-systems. *Probabilistic Engineering Mechanics*, 9:235 – 243, 1994.
133. R. Rackwitz and B. Fiessler. Structural reliability under combined load sequences. *Computers and Structures*, 9:489 – 494, 1978.
134. M. R. Rajashekhar and B. R. Ellingwood. A new look at the response surface approach for reliability analysis. *Structural Safety*, 12:205–220, 1993.

135. G. Rätsch, B. Schölkopf, A. Smola, S. Mika, T. Onoda, and K. R. Müller. Robust ensemple learning. In A. Smola, P. Bartlett, B. Schölkopf, and D. Schuurmans, editors, *Advances in Large Margin Classifiers*, pages 207 – 220. The M. I. T. Press, Cambridge, 2000.
136. A. C. Rencher. *Methods of Multivariate Analysis*. John Wiley and Sons, New York, 1995.
137. F. Riesz and S. Nagy. *Functional Analysis*. Dover Publications, New York, 1990.
138. B. D. Ripley. *Stochastic Simulation*. John Wiley and Sons, New York, 1987.
139. B. D. Ripley. *Pattern Recognition and Neural Networks*. Cambridge University Press, Cambridge, 1996.
140. C. P. Robert and G. Casella. *Monte Carlo Statistical Methods*. Springer, New York, 1999.
141. F. Rosenblatt. *Principles of Neurodynamics: Perceptron and Theory of Brain Mechanisms*. Spartan Books, Washington, 1962.
142. R. Y. Rubinstein. *Simulation and the Monte Carlo Method*. John Wiley and Sons, New York, 1981.
143. P. Ruján and M. Marchand. Computing the Bayes kernel classifier. In A. Smola, P. Bartlett, B. Schölkopf, and D. Schuurmans, editors, *Advances in Large Margin Classifiers*, pages 329 – 348. The M. I. T. Press, Cambridge, 2000.
144. J. M. F. Saraiva and N. F. F. Ebecken. Aplicaçaõ de redes neuronais artificiais no estudo da confiabilidade estrutural. *Revista Internacional de Métodos Numéricos para Cálculo y Diseño en Ingeniería*, 14:167–180, 1998.
145. T. Sasaki. A neural network-based response surface approach for computing failure probabilities. In R. B. Corotis, G. I. Schuëller, and M. Shinozuka, editors, *Structural Safety and Reliability - Proceedings of the International Conference on Structural Safety and Reliability, ICOSSAR 01, Newport Beach, California, 17-22 June 2001*, page 257, Lisse, The Netherlands, 2001. A.A.Balkema, Publishers.
146. C. A. Schenk, H. J. Pradlwarter, and G. I. Schuëller. On the dynamic stochastic response of fe models. *Probabilistic Engineering Mechanics*, 19:161 – 170, 2004.
147. M. Schetzen. *The Volterra and Wiener Theories of Nonlinear Systems*. Krieger Publishing Company, Malabar, 1989.
148. B. Schölkopf and A. Smola. *Learning with Kernels*. The M. I. T. Press, Cambridge, 2002.
149. B. Schölkopf, A. Smola, and K. R. Müller. Kernel principal component analysis. In B. Schölkopf, C. J. C. Burges, and A. Smola, editors, *Advances in Kernel Methods*, pages 327 – 352. The M. I. T. Press, Cambridge, 1999.
150. G. I. Schuëller, C. G. Bucher, U. Bourgund, and W. Ouypornprasert. On efficient computational schemes to calculate failure probabilities. *Probabilistic Engineering Mechanics*, 4:10 – 18, 1989.
151. G. I. Schuëller and R. Stix. A critical appraisal of methods to determine failure probabilities. *Structural Safety*, 4:293 – 309, 1987.
152. D. W. Scott. *Multivariate density estimation*. John Wiley and Sons, New York, 1992.
153. A. Sen and M. Srivastava. *Regression Analysis*. Springer Verlag, New York, 1990.

154. M. Shinozuka. Stochastic fields and their digital simulation. In G.I. Schuëller and M. Shinozuka, editors, *Stochastic Methods in Structural Dynamics*, pages 93 – 133. Martinus Nijhoff Publishers, Dordrecht, 1987.
155. M. Shinozuka and Y. Sato. Simulation of nonstationary random processes. *Journal of Engineering Mechanics*, 93:11–40, 1967.
156. E. Simiu and R. H. Scanlan. *Wind Effects on Structures.* John Wiley and Sons, New York, 1996.
157. K. Sobczyk and J. Trębicki. Maximum entropy principle in stochastic dynamics. *Probabilistic Engineering Mechanics*, 5:102 – 110, 1990.
158. T. T. Soong and M. Grigoriu. *Random Vibration of Structural and Mechanical Systems.* Prentice Hall, Englewood Cliffs, 1993.
159. P. D. Spanos and B. A. Zeldin. Galerkin sampling method for stochastic mechanics problems. *Journal of Engineering Mechanics*, 120:1091–1106, 1994.
160. B. F. Spencer, M. K. Sain, C. H. Won, D. C. Kaspari, and P. M. Sain. Reliability-based measures of structural control robustness. *Structural Safety*, 15:111–129, 1994.
161. A. Sternberg. Stability investigation of long-span bridges using indicial functions with oscillatory terms. In I. Elishakoff and Y. K. Lin, editors, *Stochastic Structural Dynamics 2 - New Practical Applications*, pages 128 – 145. Springer-Verlag, Berlin, 1991.
162. B. Sudret and A. DerKiureghian. Stochastic finite element methods and reliability. a state-of-the-art report. Technical report, Report UCB/SEMM-2000/08, Department of Civil and Environmental Engineering, University of California, Berkely, 2000.
163. J. A. K. Suykens, T. VanGestel, J. DeBrabanter, B. DeMoor, and J. Vandewalle. *Least Squares Support Vector Machines.* World Scientific, Singapore, 2002.
164. A. Tarantola. *Inverse Problem Theory.* Elsevier, Amsterdam, 1987.
165. G. K. Tazenotopoulos, J. Goldberg, J. J. Rushanan, and M. Hausner. Discrete mathematics. In D. Zwillinger, editor, *Standard Mathematical Tables and Formulæ*, pages 161 – 249. C.R.C. Press, Boca Ratón, 1996.
166. S. Theodoridis and K. Koutroumbas. *Pattern Recognition.* Academic Press, London, 1999.
167. A. N. Tikhonov and V. Y. Arsenin. *Solutions of Ill-Posed Problems.* V. H. Winston, Washington, 1977.
168. M. E. Tipping. The relevance vector machine. In S. A. Solla, T. K. Leen, and K. R. Müller, editors, *Advances in Neural Information Processing Systems*, page 652 658. The M. I. T. Press, Cambridge, 2000.
169. M. E. Tipping. Sparse bayesian learning and the relevance vector machine. *Journal of Machine Learning Research*, 1:211 – 244, 2001.
170. Simon Tong and Daphne Koller. Support vector machine active learning with applications to text classification. In Pat Langley, editor, *Proceedings of ICML-00, 17th International Conference on Machine Learning*, pages 999–1006, Stanford, 2000. Morgan Kaufmann Publishers, San Francisco.
171. J. Trębicki and K. Sobczyk. Maximum entropy principle and non-stationary distributions of stochastic systems. *Probabilistic Engineering Mechanics*, 11:169–178, 1996.
172. L. H. Tsoukalas and R. E. Uhrig. *Fuzzy and Neural Approaches in Engineering.* John Wiley and Sons, New York, 1997.

173. E. Vanmarcke. *Random Fields: Analysis and Synthesis.* The M. I. T. Press, Cambridge, 1983.
174. V. N. Vapnik. *Statistical Learning Theory.* John Wiley and Sons, New York, 1998.
175. V. N. Vapnik. *The Nature of Statistical Learning Theory.* Springer Verlag, New York, 2000.
176. V. N. Vapnik and A. Ja. Chervonenkis. On the uniform convergence of relative frequencies of events to thir probabilities. *Theory of Probability and its Applications*, 16:264 – 280, 1971.
177. V. N. Vapnik and A. Ja. Chervonenkis. *Teoria raspoznabania obratsov [Theory of Pattern Recognition].* Nauka, Moscow, 1974.
178. V. N. Vapnik, S. Golowich, and A. Smola. Support vector method for function approximation, regression estimation and signal processing. *Advances in Neural Information Processing Systems*, 9:281 – 287, 1996.
179. B. Vidanovic. *Statistical Modeling by Wavelets.* John Wiley and Sons, New York, 1999.
180. G. Wahba. Support vector machines, Reproducing Kernel Hilbert Spaces and randomized GACV. In B. Schölkopf, C. J. C. Burges, and A. Smola, editors, *Advances in Kernel Methods*, pages 69 – 88. The M. I. T. Press, Cambridge, 1999.
181. G. G. Walter. *Wavelets and other orthogonal systems with applications.* CRC Press, Boca Ratón, 1994.
182. E. A. Wan. *Finite Impulse Response Neural Networks With Applications In Time Series Prediction.* PhD thesis, Stanford University, 1993.
183. S. R. Winterstein. Nonlinear vibration modes for extremes and fatigue. *Journal of Engineering Mechanics*, 1114:1772–1790, 1988.
184. L. Xu and G. Cheng. Discussion on: moment methods for structural reliability. *Structural Safety*, 25:193 – 199, 2003.
185. P. V. Yee and S. Haykin. *Regularized Radial Basis Function Networks.* John Wiley and Sons, New York, 2001.
186. B. A. Zeldin and P. D. Spanos. Random field representation and synthesis using wavelet bases. *Journal of Applied Mechanics ASME*, 63:946 – 952, 1996.
187. Y. G. Zhao and T. Ono. A general procedure for first/second-order reliability method (FORM/SORM). *Structural Safety*, 21:955 – 112, 1999.
188. Y. G. Zhao and T. Ono. New approximations for SORM: Part I. *Journal of Engineering Mechanics*, 125:79–85, 1999.
189. Y. G. Zhao and T. Ono. New approximations for SORM: Part II. *Journal of Engineering Mechanics*, 125:86–93, 1999.
190. Y. G. Zhao and T. Ono. Moment methods for structural reliability. *Structural Safety*, 23:47 – 75, 2001.
191. K. Ziha. Descriptive sampling in structural safety. *Structural Safety*, 17:33 – 41, 1995.

Essential Symbols

b =	threshold of the hyperplane
c_i =	class of sample $\boldsymbol{x}_i$
$c(\boldsymbol{x})$ =	classification function of sample $\boldsymbol{x}$
C =	weight of slack variables
$\boldsymbol{C}$ =	covariance matrix
$C(t_1, t_2)$ =	covariance function of a random process
d =	number of dimensions of input space
D =	number of dimensions of feature space
$E(\cdot)$ =	expected value
$f_j(t)$ =	$j-$th eigenfunction of the covariance kernel
$\mathcal{F}$ =	failure domain
$g(\cdot)$ =	limit state function
$G(n)$ =	growth function for n samples
$h(\boldsymbol{x})$ =	component function of a non-adaptive model
$h(\boldsymbol{x}, \boldsymbol{x}_i)$ =	component function of an adaptive model
$\boldsymbol{H}$ =	Hessian matrix
$\boldsymbol{I}$ =	identity matrix
K =	sum of principal curvatures
$K(\boldsymbol{z}, \boldsymbol{y})$ =	kernel function
$\boldsymbol{K}$ =	Gram matrix derived from the kernel function
$L(\cdot, \cdot)$ =	cost function
n =	number of samples
N_{L} =	sample complexity using algorithm L
$\mathbb{N}$ =	set of natural numbers
$p_{\underline{\boldsymbol{x}}}(\boldsymbol{x})$ =	joint density function of variables $\underline{\boldsymbol{x}}$
$p(\boldsymbol{z}\|\mathcal{F})$ =	conditional class-$\mathcal{F}$ density function
$p(\boldsymbol{z}\|\mathcal{S})$ =	conditional class-$\mathcal{S}$ density function
P_{f} =	probability of failure
$P(\mathcal{F})$ =	prior probability of class $\mathcal{F}$
$P(\mathcal{S})$ =	prior probability of class $\mathcal{S}$

R	=	average radius of curvature
$R(\boldsymbol{\omega})$	=	risk function
$R_{\mathrm{e}}(\boldsymbol{\omega}, n)$	=	empirical risk with n samples
$R_{\mathrm{r}}(\boldsymbol{\omega}, n)$	=	regularization risk with n samples
$\mathbb{R}$	=	set of real numbers
s	=	number of support vectors
$\mathcal{S}$	=	safe domain
v	=	Vapnik-Chervonenkis dimension
v_j	=	principal component of order j
V_0	=	fundamental volume of the safe domain
$\underline{x}$	=	basic random variable
$\underline{\boldsymbol{x}}$	=	vector of basic random variables
$\underline{u}$	=	normalized random variable
$\underline{\boldsymbol{u}}$	=	vector of normalized random variables
$\boldsymbol{w}$	=	weight vector in a linear combination of functions
α_i	=	Lagrange multiplier of vector i
$\boldsymbol{\alpha}$	=	vector of Lagrange multipliers
β	=	reliability index
Δ	=	normalized maximum margin in the support vector method
Δ_i	=	margin of sample i
ϵ	=	regression error
ζ_i	=	basic variable in the feature space
$\boldsymbol{\zeta}(\boldsymbol{x})$	=	nonlinear mapping
θ	=	parameter of kernel functions
λ_j	=	$j-$eigenvalue of the covariance kernel
$\boldsymbol{\mu}$	=	mean vector
μ	=	mean
ξ	=	slack variable for non-separable problems
ρ	=	radius of smallest sphere containing the samples in feature space
ρ_{ij}	=	correlation coefficient betwen variables i and j
σ	=	standard deviation
$\Upsilon(A)$	=	indicator function of statement A
$\phi(x)$	=	standard Gaussian density function of x
$\phi(t)$	=	wavelet scaling function
$\Phi(x)$	=	standard Gaussian distribution function of x
$\psi(t)$	=	wavelet function
$\boldsymbol{\omega}$	=	set of parameters of a family of approximating functions $h(\boldsymbol{x}, \boldsymbol{\omega})$

Index

Printing and Binding: Strauss GmbH, Mörlenbach